THE TIME DIMENSION:

ITS RELATIONSHIP TO THE ORIGIN OF LIFE

THE TIME DIMENSION:
ITS RELATIONSHIP TO THE ORIGIN OF LIFE

Dr. A. E. Wilder-Smith

THE WORD
FOR TODAY
P.O. Box 8000 Costa Mesa CA 92628

The Time Dimension: Its Relationship to the Origin of Life

Published by: The Word For Today Publishers
P. O. Box 8000 Costa Mesa CA 92628 (800) 272–WORD
ISBN# 0–936728–45–0

Table of Contents

Preface....1

Chapter I

The Scientific Plausibility of a Naturalistic Approach To Life's Origin....5

Fox And Miller....5
The Origin Of All Machines....6
The Von Neumann Machine....7
Information Is Required To Instruct All Machine Construction....8
Spontaneous Generation: A Philosophy – Not A Science....8
For Biogenesis Optical Purity Is Mandatory....9
The Importance Of Teleonomy....9
The Second Law Of Thermodynamics....10
The DNA Molecule....10
The Nature And Significance Of Optical Activity And Molecular Asymmetry....11
The Oxford Union Debate And The Juelich (Germany) Symposium....19
The Science Of Chemistry Denies....23
Entropy Gradients....23
The DNA Three – Dimensional Information....25
Actual Information Mandatory For....26
The Claude Shannon Type Of Information....27
Louis Pasteur....27
Arthur Kornberg And Sol Spiegelman....28
The Grand Problem Of Biogenesis Today....31
Origin Of Actual Information....32

Chapter II

The Nature Of Information....35

I) The History and Nature of....35
An Alternative In Information Theory –....39
The Binary Digit....40
Words, Letters, Sequences and Meaning....40
An Example....41
The Work Of Professor Werner Gitt....45
Professor Gitt Continues....47
Actual and Potential Information –....52
Fox And Miller's Philosophy Summed Up....54
The Role of the Cerebrum in the Production of Actual Conceptual Information....61

The Possibility Of A Universal Think–Tank? 63

Chapter III

The Space–Time Continuum 67
b) The Nature of the Space–Time Dimension 69
Multidimensional Reality 71
New Dimensions in Space, Time and Space–Time 72
Views On The Nature Of Time (Leibnitz) 73
Views On The Nature Of Time (Sir Isaac Newton) 74
The Nature of Thought and Computers 75
Jacques Monod's Materialism 77
The Psychospace of the Mind and its Conditioning 78
The Substance of Thought 79
The Nature of the Psychospace 80
The Time Dimension is Not Creative 80

Chapter IV

The Theoretical And Experimental Basis Of Geological And Evolutionary Dating 85
Creation and Measuring the Time Dimension 85
Long Term Dating Methods 88
Experiments With Long Term Dating Calibration In Iceland 90
No Uncalibrated Long Term Radiometric Dating Method To Be Taken Seriously 91
The C14 Method Of Dating 94
The Lead Isotope Method of Dating 94

Chapter V

Some Recent Fossil Evidence 97
1) Fresh Hadrosaur Fossils in Alaska 97
Comment on Basinger's Above Statements 104
Some of the Costs of the Darwinian Theory to the Taxpayer 107
Addendum 110
Mirror Images and Entropy Status 111
Enzymes, the basis of the biological procurement of metabolic energy 112

The Time Dimension: Its Relationship To The Origin Of Life

Preface

By the time the average child has completed the school system and by the time the average undergraduate has completed college or university he feels himself sure of the following five "scientific" facts:

1) The universe and the earth are billions of years old, which is the time required for inorganic material to be transformed spontaneously into living biological material and the time required by the primeval cell to evolve spontaneously and by natural selection up to homo sapiens sapiens. Not only the schools, colleges and universities are full of this "information" but also magazines like "The National Geographic" can scarcely write a paragraph without citing this sort of 'information', directly or indirectly, as a scientific fact.

2) The evolution of life from non–life and the evolution of the primeval amoeba–like cell up to homo sapiens has allegedly taken millions of years too. Chance mutations followed by natural selection are slow processes – unless our undergraduate happened to be a student of Dr. Stephan J. Gould who thinks a little differently on this subject although he professes to be a Darwinian evolutionist.

3) There is no longer any need to believe any more in a designer to account for the manifest design in biology. Mutation followed by natural selection will produce results

which look like design, without the presence of any intelligence behind the design so produced – although they both need long time periods to achieve the result.

4) The majority of great scientists who were convinced Christians in the past and believed the biblical creation report and time table, were so because the science they knew about was lacking in maturity. That is, in plain language they were Christians because they did not know the whole truth that modern science has allegedly taught us.

5) Really advanced science would deny the need for believing in any personal intelligent Creator.

The following treatise examines in detail the above five points in the next five chapters. It shows that not one of their concepts is founded on science carried out according to the well known rules of scientific procedure in the laboratory today. If the scientist looks really carefully into the methods used to establish the five points mentioned above he will find, as we show in the following text, that even the old age of the earth (billions of years) and the universe rests upon excellent mathematics but which is unfortunately for mainstream biology *completely erroneously applied*.

To be sure of treating the subject of biogenesis, time and evolution effectively the author has had to look into such subjects as the nature of time, the nature of optical asymmetry in chemistry, the measuring techniques of dating and certain recent fossil finds which are not yet generally known, but which bear on the subject of this book.

The radiometric instrumentation used for calculating the age of the earth in billions of years is known to be completely uncalibrated and certainly not an established "fact" of modern science but an established cause of error. The use of *uncalibrated experimental tools* has always been a cardinal error of all really scientific method.

The reason why Darwin and so many other scientists lost their faith in their earlier joyous specifically, Christian convictions, lies neither in the facts of Genesis nor in the creation account being erroneous but in the fact that science has often thought too superficially about *its experimental methods and has therefore drawn erroneous conclusions from them.*

The following treatise examines the above 5 points critically. All the author requests of his readers is a fair unbiased but cold examination of what experimental science has to say today on matters which bear as a whole on the biblical testimony in the laboratory.

Chapter 1
The Scientific Plausibility of a Naturalistic Approach To Life's Origin

Mainstream science in the universities, places of higher learning and high schools throughout the modern world has taught as though it were a scientific fact for years now that life arose from inorganic matter and a primeval "slime". This occurred by the spontaneous processes of simple organic chemistry working on the products synthesized by the interaction of electrical energy on mixtures of gases such as methane, ammonia and steam in the earth's atmosphere over immense periods of time.

FOX AND MILLER

Fox and Miller showed many years ago now that mixtures of amino acids are indeed produced by the passage of dark or bright electrical discharges through methane, ammonia and steam and that such mixtures contained some of the amino acids which figure in the basic constitution of material life. [1]

---

1 (S.L. Miller: *Science*, 117, (1953) 528. See also: A.E. Wilder-Smith, "*The Natural Sciences know nothing of Evolution*", T.W.F.T. Publishers, Costa Mesa, Ca. 92628, P.O. Box 8000 U.S.A. p. 11. cf. A.E. Wilder-Smith, "*Die Naturwissenschaften kennen keine Evolution*", Schwabe & Cie, Basel, Switzerland, editions of 1978, 1978, 1980, 1982, 1985.

It was therefore concluded by most biological scientists that lightning passing through the earth's early atmosphere yielded the organic chemicals required for life's origin. Life supposedly arose spontaneously over extremely long time periods from these racemic amino acids and other substances.

The chemical properties of inorganic matter aided by the immense periods of time which astronomers and others attribute to the age of matter, allegedly did the rest and effected the spontaneous synthesis of life from these spontaneously formed amino acids.

Fox and Miller and many others with them, concluded that this electrochemical origin of certain of the amino acids required for the synthesis of the biological cell represented concrete evidence for the purely naturalistic origin of the biological cell.

Mainstream scientific circles have believed ever since this work by Fox and Miller that the phenomenon of life on planet earth owes its origin solely to electrochemical processes, spontaneous organic reactions and time and has nothing to do with an origin in intelligence or personal will. Life is, therefore, thought to be the product of organic chemistry, immense periods of time, electricity, inorganic chemicals and of nothing else.

THE ORIGIN OF ALL MACHINES

No personal creator working specifically with matter and with the intention of synthesizing life from inorganic matter is, in this view, necessary to explain the machine phenomenon of life: allegedly, the whole metabolic machine of life arose from the properties of matter alone (which some original creator may, of course, have endowed matter with) working over immense periods of time. In this particular scenario we have the suggestion that the most complex machine in the universe, namely life itself, arose by chance processes— *the first machine in fact of any sort ever known to have arisen by chance processes plus the inherent properties of matter and not by the voluntary interaction of a personal reason and will with matter. This would make the biological machine the first machine not to owe its origin to personal will and intention but to pure randomness together with the inherent properties of matter.*

THE VON NEUMANN MACHINE

Yet at the same time the most complex machine known to man, – the living metabolic life machine is, as is well known, *a von Neumann machine,* the component parts of which *repair themselves* after moderate injury. Most man–made machines cannot effect this feat of self–repair or self replication.

It would appear very difficult to seriously believe as a scientist that any machine arose by random processes plus the inherent properties of matter but far more difficult still to believe that a self–repairing, self–replicating von Neumann machine arose by such chance processes too! For those readers who are not familiar with the von Neumann Machine, I have described such in some of my previous books on evolutionary theory . [2]

Scientists who have recognized that a scenario of the above Darwinian kind will not do to explain the origin of a von Neumann machine such as life is, have often come to the conclusion that therefore some extra factor, maybe information, does after all, play a role in the mechanism of the origin of a von Neumann life machine. On the other hand, the idea that such necessary information might point to an origin

---

[2] *loc. cit.* p. 5

of life founded in a phenomenon known as personality or will, is anathema to the modern "Zeitgeist".

INFORMATION IS REQUIRED TO INSTRUCT ALL MACHINE CONSTRUCTION

Therefore, information experts maintain that even the information and instructions necessary for biogenesis must have arisen by chance to avoid the necessity of postulating a personal Creator. An idea like that takes fire immediately, for it fully satisfies the modern Zeitgeist. Scientists only rebel about introducing the idea of information as a necessary ingredient in the origin of life when it comes to the idea of a Creator who might be a Person and personally interested in the project of life and therefore directly concerned in initiating biogenesis by combining actual information with matter. [3]

Scientists of most shades of opinion have nothing against the concept that information is necessary for biogenesis and is stored as such on the DNA molecule – *provided the information concerned is formed by chance processes as well, and not by personal will. That is, provided the information is not coupled in its origin to thoughtful personal processes but only to chance.*

SPONTANEOUS GENERATION: A PHILOSOPHY – NOT A SCIENCE

Views leaning to the idea of spontaneous generation are, of course, philosophical speculation and nothing else and have little to do with experimental science. The remarkable fact is that such purely philosophical views, (attributing biogenesis to stochastic spontaneous processes only) although they have little foundation in experimental science, are without any questioning accepted by the present scientific world and then

---

[3] See my "*The Natural Sciences Know Nothing of Evolution*" and my "*Die Naturwissenschaften kennen keine Evolution*"

taught as *experimentally* factual and therefore as scientific fact. It is totally forgotten that the philosophical half truths involved are thus being seen as total experimental truths.

FOR BIOGENESIS OPTICAL PURITY IS MANDATORY

It is true that: 1) certain racemic amino acids are produced by stochastic methods (Fox and Miller). *It is forgotten, however, that the other half of the truth must be reckoned with too, namely that:* 2) life and living cells cannot be synthesized from racemic amino acids or from any other racemic components. Theory as well as experiment teach us very plainly just this lesson as we shall now demonstrate.

The properties of matter, the laws of chemistry together with the properties of electricity working over immense spans of time allegedly explain in this purely philosophically based view everything there is to explain concerning the origin of life.

The purely scientific evidence against the above mainstream philosophy on the origin of life is seldom presented in the scientific research journals or text books today so that we propose to present at least some of it in the following chapters.

THE IMPORTANCE OF TELEONOMY

The grand difficulty of any purely naturalistic scenario for life's origin and one which is seldom mentioned in the literature, is the fact that inorganic matter left to itself never shows any sign of the resident teleonomy which would be required to guide the synthesis of matter into the fabulously complex von Neumann machine which life is.

THE SECOND LAW OF THERMODYNAMICS

The Second Law of Thermodynamics states that matter tends to thermodynamic equilibrium and therefore not to increasing complexity. (see Fig. 6 p. 81)

This law makes the whole Darwinian naturalistic scheme of things difficult to swallow scientifically. Modern biological research has conclusively demonstrated that the teleonomy required to guide the synthesis of matter into living cells resides on the DNA molecule (and partially at least on certain protein molecules too) which stores the teleonomy required as very concrete and definable bits of information.

THE DNA MOLECULE :THE KINGPIN FOR BIOGENESIS

Modern science has conducted an astounding piece of "question begging" by assuming that inorganic matter is itself the real and original seat of the storage of this biological teleonomic information. That is, that inorganic matter itself rather than that a particular aggregate of matter (the DNA or other molecule) exhibits the capacity for storing the information necessary for biogenesis and for evolutionary biology which then evolves itself up to life given sufficient spans of time.

The concept that spontaneous chemical reactions working over huge periods of time conspired to synthesize the first living cell is certainly no fact of science but rather one of purest fiction. It is, however, a fact of science that the bits of teleonomic information stored on the DNA molecule have never been experimentally or scientifically replaced by any information stored on inorganic matter at large to generate any known form of life.

The grand question of biogenesis is chemical in nature and concerns the origin of the molecular asymmetry leading to optical purity which living matter shows. Naturalistic processes cannot lead to the optical purity necessary for life to start up because there is no entropy gradient between the levo and dextro isomers which means that common or garden chemistry cannot separate such optical isomers without specific information being supplied from outside the system to perform this feat of pattern recognition.

The scientific problem of problems in all questions of biogenesis is really: is a spontaneous origin of the optically active DNA molecule of biology together with its actual information store scientifically feasible or experimentally founded? One can extend this question to the wider one: is the genesis of any optically pure molecule such as we find in large quantities in all forms of life, at all attributable to spontaneous, purely naturalistic chemical reactions?

THE NATURE AND SIGNIFICANCE OF OPTICAL ACTIVITY AND MOLECULAR ASYMMETRY

The biological metabolic machine of life is constructed on the basis of "molecular fit" which principle is rather difficult to conceive of for those who have not had the advantage of taking a course of advanced organic chemistry. But the idea can fairly easily be made plain to anyone who will listen long enough!

The biological molecules which make the physical basis of life possible need to react with one another in order to provide energy for the living cell. They must be brought very close to one another in order to react so that the molecular distance between the two reacting molecules becomes very small indeed. To achieve this close proximity between the reacting molecules, the whole biological cell system is built on the principle of a hand fitting into a glove. Drugs react with the body by the identical principle. There are so–called receptor sites which act like a glove into which the drug slips and thus gains close proximity to the machinery of the cell. But the fit must be well nigh perfect to achieve this end. Such molecules are called receptor sites. The reacting molecule slips into the receptor site like a hand slipping into its glove.

Such a system of “fit” brings with it very close proximity between the two substances permitting the proximity necessary for chemical reaction at low temperature.

It will be obvious that it is just as difficult to make a left handed glove as it is to make a right handed one. The scientists express this state of affairs by saying that the "entropy status"of the left handed molecule is exactly the same as that of the right handed molecule. The left handed glove is exactly as easily made as the right handed one. (See Fig. 1 p. 13)

This state of affairs is clarified by noting that the left handed molecule rotates the plane of any polarized light passed through it mostly to the left and the right handed molecule rotates it to an equal extent in the opposite direction.

CH4 + H2O + NH3
e (energy)
CH3–CH–COOH
NH2
ALANINE (Racemate)
Three Dimensional
Tetrahedrons
CH3
CH
NH2
COOH
CH3
CH
COOH
NH2
MIRROR IMAGES
No Entropy Gradient
between mirror image
& its original

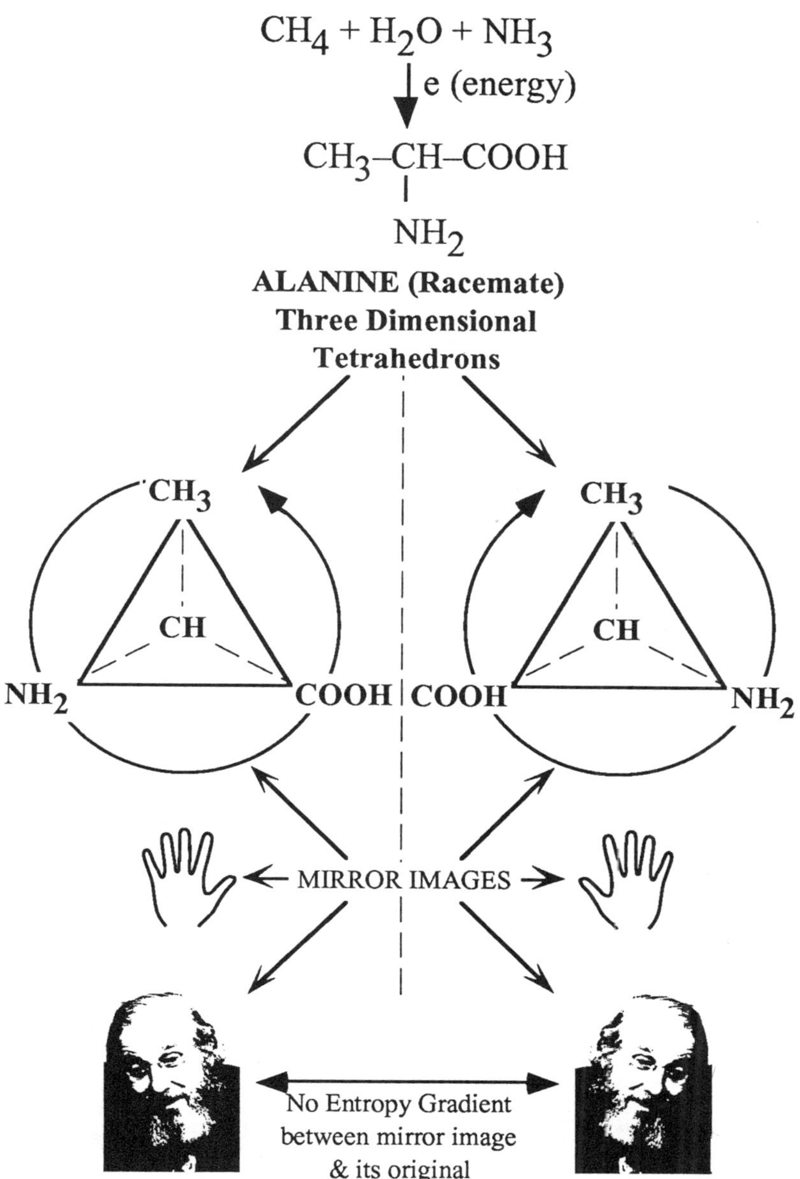

Figure 1

FORMULA: CREATION and EVOLUTION

EVOLUTION:

1) m + t + e = abiogenesis

2) Primal cell + t + e + mutation + natural selection = evolutive speciation

CREATION:

3) m + t + e + i = abiogenesis

4) cell + t + e + i = homo sapiens

m = matter t = time e = energy i = actual information

Figure 1a

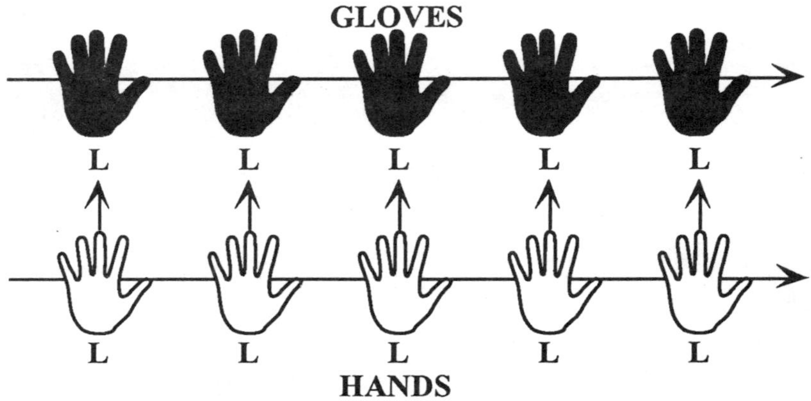

Perfect fit of hand–glove mechanism only when all 'hands' are of identical 'handedness', i.e. lefthanded

Figure 2

ENERGY PROCUREMENT in the BIOLOGICAL CELL

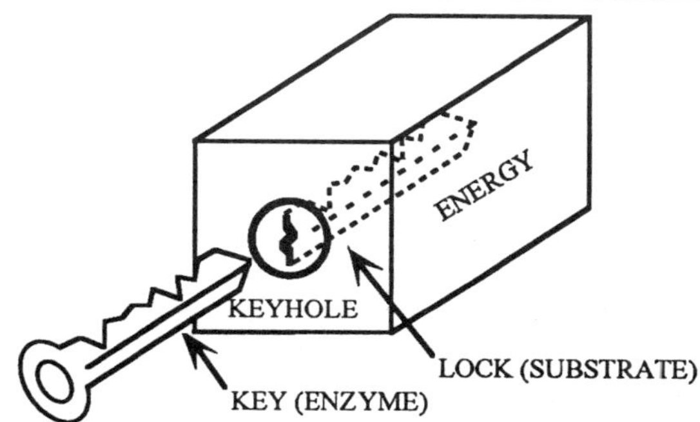

TWO METHODS of OPENING "SAFE"
1] Acetylene Torch + Oxygen
2] Lock & Key Mechanism

Figure 3

How does the hand–glove system arise in chemistry? It is perfectly simple when one takes into account the basic structure of the carbon atom which is the element of which life is principally constructed.

The carbon atom takes the shape of a tetrahedron: Its form resembles that of four equilateral triangles so laid together that the four points of the tetrahedron point outwards into space. (See Fig. 1 p. 13)

If now the substituents on the four points of the carbon tetrahedron are all different, then, as appears in Fig. 1, there are produced automatically as a mere fact of geometry two kinds of molecules which are mirror images. One of these mirror images corresponds to the left handed glove we have been discussing and the other corresponds to the right handed one.

It is an observation of science that if the left handed isomer or glove is dissolved in a solvent and polarized light is passed through it, then the plane of the light will be tilted say to the left. And if light is passed through a solution of the right–handed isomer then the plane of the light will be tilted in the opposite direction. One mirror image of the tetrahedron represents one type of glove and the other mirror image represents the opposite type of glove.

The left handed glove will, however, only fit the left hand and the right handed one will only fit the right hand. If these "fits" are not exact, no metabolism or chemical reactions can take place. No life is possible without the "handedness" being correct. Organic chemistry alone without external actual information cannot arrange for this "fit" so that no metabolism for life's energy supply could possibly be arranged without the actual information needed to effect "hand–glove" arrangements for biogenesis to take place. One may use the analogy of a key fitting into a lock to express the hand–glove idea equally well.

This is the basis of metabolism in all biology. By this method only are the reacting substances of all biology brought closely enough together to react with one another at the relatively low temperature of the body. That is why biology is able to "burn" sugar and other substances at such low temperatures. The molecules are brought so close together by the hand–glove fit (or the key–lock mechanism) method, that they react at low temperature to supply the energy needed for life to work.

For this reason life could not start at any temperature unless the "hand–glove fit" (key–lock) technique had been developed first. And precisely that fact requires the chemical optical isomerism to be in place before life could get started. And that again requires actual information to enable chemistry to separate the left handed isomers from the right handed ones to ensure perfect chemical "fit" – otherwise chemical metabolism would not be possible. For closeness chemically speaking (proximity) is the first requirement of chemical metabolism and reaction.

Before life could start at biogenesis, chemistry must have been assisted by the presence of suitable actual information to help chemistry overcome the fact that the lack of an entropy gradient between mirror images prevented chemistry from providing the close fit needed to start metabolism and to provide energy for life's functions. No Shannon type of information (sic) is capable of satisfying this requirement of energy supply.

If now the supply of substances for life's synthesis consists of a mixture of left handed and right handed amino acids (to take just an example of substances for biogenetical synthesis) for building say enzymatic proteins, then the resulting sequences of amino acids in the protein formed will look like Figure 4 p. 18 shows. That is there will result a "higglede–piggledy" of left and right sequences which will not slot into any receptor site on the cell, for the sequences are chaotic.

The net result will be that no metabolism to obtain energy will result for there is no possible "fit" of d or l sequences. The whole sequence of d and l in the biological protein molecules must fit into the d and l sequences present on the acceptor molecule. Just one wrong sequence of a d molecule in a whole chain will render the fit of any acceptor–receptor system invalid. (See Fig. 4 p. 18) This means that optical purity in synthesis of this type must be perfect if a living metabolizing biological cell is to result.

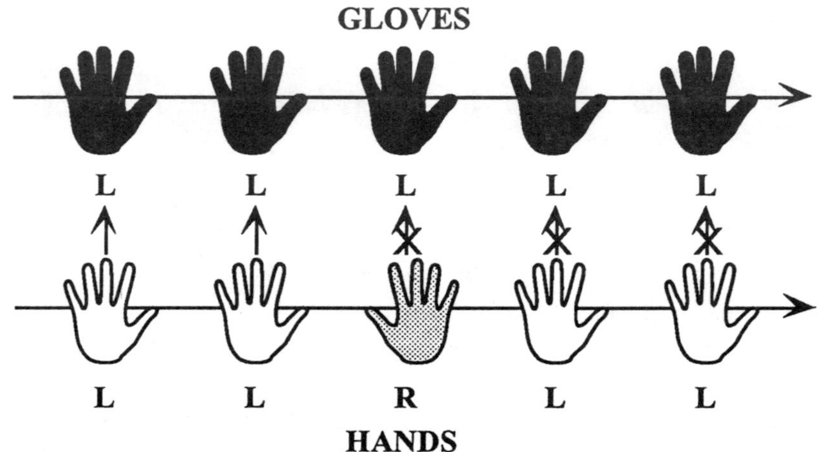

If the least optical impurity (wrong–handedness) is introduced, the entire reaction comes to an abrupt halt, because each link in the chain is dependant upon the fit of the entire molecule. ⸸ = no "fit"

Figure 4

If the system in figure 4 is regarded as a zipper–fastener mechanism, then when the "zipper" comes to the wrong dextro or levo isomer, the mechanism will jam, and no metabolism will occur.

The rather abstruse chemical and geometrical facts can be equally well illustrated by comparing the d and l isomers in a protein sequence to notches on a key in a lock and key mechanism. (See Fig. 3 p. 15)

If there is a sequence of notches on a key to make it fit a lock, just one notch sequence which is wrong, will invalidate the total fit. Thus the starting optical purity for substances to begin life must be perfect. Naturalistic chemistry can supply only a racemate because there is no entropy gradient between the dextro and levo isomers. Therefore actual information is required to steer biogenetical chemistry into the required dextro or levo isomer. No stochastic reactions can supply this kind of actual information, so that naturalistic chemistry [that means chemistry left to itself] cannot supply the optically pure starting materials mandatory to start life.

This is the reason that there is *no scientific basis for looking for spontaneously formed life on the outer planets and galaxies.* If there is such life out there, it will have to have been synthesized by the combination of actual information with matter just as all life on planet earth must have arisen. That is, if there is life "out there" it will have been synthesized by a supplier of actual information, that is by a Creator, just as is the case on planet earth. In any case such life "out there" will not be a proof that Darwinian Evolution is valid for biogenesis in the whole universe, as most evolutionists wish us to believe.

If present life is fed racemates to metabolize and live on, it will first break the racemate down and then re–synthesize it to pure optically active forms using the actual information on its DNA molecule to do so.

THE OXFORD UNION DEBATE AND THE JUELICH (GERMANY) SYMPOSIUM

The Oxford Union conducted on February 14th. 1986 a debate to decide whether a spontaneous naturalistic origin of life as we know it was possible on a purely scientific basis. For without spontaneous biogenesis being a fact, *naturalistic evolution itself is pointless and impossible. For if science points to a Creator having made the first cell, then why should He not have created the various species of life too after the first cell had appeared by spontaneous processes? To show scientifically that biogenesis needs the activity of a Creator and that biogenesis never occurs spontaneously is to show at the same time that there is no scientific reason for not assuming the activity of a Creator as far as the creation of species too is concerned.*

The present treatise arose partially as a result of the above mentioned debate at Oxford in which I was invited to take part. The motion before the House was formulated in the Oxford Union committee as follows: "*The Doctrine of Creation is more valid than the Theory of Evolution*".

Professor Andrews of London University and Professor Wilder–Smith formerly of the University of Illinois Medical Center in Chicago, U.S.A., both gave their scientific reasons for accepting the doctrine of creation as a *valid scientific explanation of the origin and the various species of life* rather than that of the purely naturalistic Darwinian thesis.

Several students of the University of Oxford gave their scientific reasons for accepting the concept of creation as the valid scientific explanation of life as we see it on earth now.

Before the debate commenced it was agreed in committee in the Oxford Union's President's office that no religious or non–scientific, non–repeatable material should be introduced into the debate. Only repeatable falsifiable scientific fact would be acceptable. To this point of policy the representatives of both sides of the House readily and specifically agreed.

Since the debate was never published – most Oxford Union debates are given *automatically nationwide publicity in the press and television as well as on the radio,* there may well be some cogent reason for the total cover–up which the whole debate subsequently experienced.

One could perhaps be pardoned for surmising that the fact that the creationists won some 114 of the votes from the voting public of about 300 was a matter of considerable shame for the Oxford Union which, of course, represented the purely materialistic naturalistic evolutionary viewpoint of biogenesis.

About four to six hundred guests were also present at the debate, but only some three hundred members of the Oxford Union who were present exercised voting rights.

A possible reason for the total cover–up of the Oxford Union debate is, maybe, illuminated by Richard Dawkins' impassioned plea to the audience before the voting took place and after the debate itself was over. Dawkins implored (the word he himself used) the voting public not to give a single vote for the creationist position, for every vote in favor of creationism would, he maintained, be a blot on the escutcheon of the ancient University of Oxford.

After the debate was over Richard Dawkins attacked my position, not on the basis of the scientific fact I had cited *but on the basis of my religious beliefs*. Allegedly I was a fundamentalist.

Since it had been agreed not to let religious factors play any role in the proceedings, Professor Andrews brought up the point of order, that no religious considerations should play any role. The president supported Professor Andrews and Richard Dawkins sat down.

Professor Maynard Smith then stood up and said he was glad that I had stuck to pure science in the debate, science which was impeccable, but said that I believed in a small tribal God, which was not acceptable today. He and his friends believed that the whole, big universe was God which was a superior belief to mine. *Again, I was attacked not on scientific but on purely religious grounds, which was entirely out of order.*

Subsequent efforts on the part of a librarian employed by the University of Oxford to obtain from the Oxford Union my address and a report on the debate were answered to the effect that it knew of no such debate ever having taken place and could give no information as to my person or even my present address. Thus I was obliged to send to the librarian concerned a photocopy of the invitation which the Oxford Union had sent to my correct address in Switzerland and which has in the meantime never changed, together with their formulation of the title of the motion before the House. The librarian obtained my address from friends in Australia as it was not forthcoming from Oxford.

These extreme measures taken to suppress any spread of the *scientific knowledge behind the debate and of its content, knowledge which denies any spontaneous generation of life and which denies life's naturalistic evolution from the amoeba to homo sapiens from inorganic matter spontaneously without outside help, is maybe an indication of the lengths to which some scientists will go even today. They feel obliged to suppress the spread of any knowledge of the scientific inadequacy of the materialistic atheistic naturalistic position on the origin of life. How evolutionary materialism is supported is well illustrated by the above cited facts.*

The naturalistic, Darwinian view of the origin and evolution of life is sustained in our schools and places of higher learning to a quite considerable extent not by the active spread of scientific learning and knowledge but often by its specific suppression. How much then, we may well ask ourselves, has the atmosphere in the learned modern world radically changed since the days of the Inquisition? In the present case it is *scientific fiction which is suppressing science and not the Church suppressing science.* This fact becomes particularly manifest if one tries to obtain grant money for research in exposing the present materialistic position in science. The Church used in olden days to suppress science, now it is a case of "science" suppressing science.

The Science Of Chemistry Denies The Materialistic View Of Life

For the materialistic view of the origin of life and its so–called evolution up to homo sapiens to be scientifically feasible *it must at the outset explain the origin of the optically pure asymmetric isomerism of the DNA molecule itself as well as the origin of the optically active proteins of all life. It is a well known fact of organic chemistry that organic chemistry left to itself can never produce any molecular asymmetry resulting in optical purity in any chemical compound.*

For life's metabolism is dependent on separating mirror image compounds like certain amino acids and other substances into their left handed forms and using these exclusively to build up molecules of a certain shape, like the double helix of the DNA molecule so that information can not only be stored linearly on the helices but also in depth on the spirals three dimensionally.

Entropy Gradients

As every organic chemist knows, unassisted organic chemistry is not in a position to separate these mirror images *because there is no entropy gradient between the left handed isomers and the right handed ones.* Chemistry needs the *actual information stored on the DNA molecule to effect this complete separation of the left handed mirror images from the right handed ones.* Information to effect this feat of chemistry has never been known to arise from any random processes.

The scientific reason for this is that all optical isomers show in any pair of such isomers an identical entropy status so that chemical reactions themselves cannot distinguish between say a levo isomer and a dextro one[4] although physiological reactions can certainly distinguish between such d and l optical isomers, unassisted chemistry cannot. Therefore, unassisted organic reactions can, on theoretical and experimental grounds produce only a 50% mixture of l and d isomers, which mixtures are, of course, optically inactive and useless for biogenetical synthesis.

That is, unassisted chemistry can on theoretical and experimental grounds produce only racemic mixtures of nearly equal proportions of dextro and levo isomers and no optical purity at all. Yet almost all biology is, chemically speaking, optically active, so that ordinary chemistry unassisted by actual information could not possibly have synthesized even the optically active DNA molecule nor any optically active enzymes which together start life.

That is, if no living cell was present at the beginning to start life, neither evolution nor natural selection could get going at all. Which means that without the spontaneous generation of the first living cell, there could be no Darwinian type of evolution of that cell. That is, evolution stands or falls on the possibility of spontaneous generation from inorganic matter having taken place. This cardinal fact needs scientific explanation. The whole naturalistic Darwinian viewpoint of biogenesis collapses on this one vital point of scientific fact alone, that is on the question of whether spontaneous generation occurs.

---

[4] A. E. Wilder-Smith, "*The Natural Sciences know nothing of Evolution*", TWFT Publishers, Costa Mesa, P.O. Box 8000, CA 92628, U.S.A., pp. 1, 24, 28, 30, 31, 32, 33, 37, 41, 89, 105, and also "*Die Naturwissenschaften kennen keine Evolution*", Schwabe & Cie, Basel, Switzerland, editions of 1978, 1978, 1980, 1982, 1985

THE DNA THREE – DIMENSIONAL INFORMATION STORAGE AND RETRIEVAL SYSTEM

Since, however, the DNA molecule stores its information in three dimensions rather than in merely two, and unassisted chemistry cannot build the optically pure double helix DNA spiral structure from the racemic mixtures delivered by unassisted chemistry left to its own devices and electrical discharges according to Fox and Miller, nothing can supply the optical purity required for the vital DNA molecule which cannot then be synthesized.

The third dimension characterized by optical purity allows the biological DNA molecule to store information in three rather than in two (linear) dimensions. Thus, information which linearly stored would require yards of molecular length can, by using the extra dimension supplied by optical purity and the construction of the double helix molecule, be effectively stored in micrometers of molecular length instead of in yards.

This all adds up to the fact that it is theoretically and experimentally impossible to synthesize the optically pure double helix molecule – or indeed any optically pure protein molecule either – on an *unassisted spontaneous purely chemical basis. The information to guide the DNA or indeed any other optically pure molecule into optical asymmetry must be supplied from outside the purely chemical system.* Exactly similar considerations apply to the synthesis of the other optically active proteins of biology.

ACTUAL INFORMATION MANDATORY FOR THE GENERATION OF LIFE

These facts of chemistry bring us to the necessity of recognizing that an additional factor besides chemistry alone for the synthesis of any kinds of life is mandatory which the naturalistic scenario has never seriously addressed – but which was thoroughly aired at the Oxford Union 1986 debate as well as at the *International Symposium on Generation and Amplification of Asymmetry in Chemical Systems held at Jülich in Germany on September 24th to 26th. 1973.*

The inability of plain unassisted spontaneous organic chemistry to synthesize optical purity without the help of actual information in reality destroys the whole notion of Darwinian evolution by natural selection and naturalistic processes exclusively.

Both the Oxford Union Debate and the International Symposium at Jülich confirmed that unassisted purely naturalistically guided organic chemistry is not in a position to supply the optically pure DNA molecule (or even the optically pure protein molecules which figure so largely in all biology). But both types of optically pure molecules, which are vital to the genesis and functioning of life and the three dimensional DNA storage of the prodigious amounts of information are required for the genesis and functioning of all biology. This fact of experimental science alone destroys the whole basis of materialistic Darwinism – a fact which has been covered up by atheistic scientists for years now who well know its importance and therefore suppress the information.

At the risk of being tedious it is necessary to emphasize and reemphasize the above facts, for in some general as well as in many specialist circles they are too often inadequately dealt with, or indeed effectively suppressed, as was the Oxford Union Debate on the subject mentioned in the foregoing text.

Unassisted chemistry (that is chemistry unassisted by *actual information cannot on well known theoretical and experimental grounds lead to the spontaneous generation of life. Which fact has been borne out by every experiment carried out up to the present day.* [5]

THE CLAUDE SHANNON TYPE OF INFORMATION

The standard Claude Shannon type of information (measured in bits and bytes) has theoretically and practically little in common with the *actual information* needed to "finance" biological processes, only *actual information* can supply the optically pure molecule needed for biogenesis. [6]

Thus *actual information is a scientific prerequisite for the genesis of any form of biology at present known to man – if only to supply the optical purity required for life's chemistry and information storage and retrieval system.*

LOUIS PASTEUR

Even a century ago Louis Pasteur recognized this fact and instituted a special program of research aimed at generating life and therefore optically active molecules by applying magnetic fields to his reaction mixtures.

Pasteur found that his program was all in vain, for chemistry and matter needed *other assistance besides magnetism to generate optical purity, assistance, in fact, of an hitherto unsuspected type which has since been brilliantly worked out by men such as Arthur Kornberg and Sol*

---

[5] A.E.Wilder-Smith, "*The Scientific Alternative to Neo-Darwinian Evolutionary Theory*", p. 33 (new edition), TWFT Publishers, Costa Mesa, P.O. Box 8OOO, CA 92628

[6] "*The Scientific Alternative to Neo-Darwinian Evolutionary Theory*" p. 33 (New edition), "*The Natural Sciences know Nothing of Evolution*", pp. 33-41 (New edition), TWFT Publishers, P.O. Box 8000 Costa Mesa, CA 92628, U.S.A. on Information Generation

Spiegelman in their successful syntheses of certain self replicating phages from inorganic matter. [7]

ARTHUR KORNBERG AND SOL SPIEGELMAN

These two scientists (Arthur Kornberg and Sol Spiegelman) used *the physical shapes* of certain living viruses to supply the *actual information* concerning shape, and therefore form which inorganic matter needed before it could spring into life. They utilized their own chemically oriented skills in applying these shapes and forms (in this case *actual information*) to inorganic matter until the shapes of the components were just correct to allow the resultant component parts to function as metabolic machines do.

To express this feat of genius on the part of Sol Spiegelman and Arthur Kornberg[8] in language which is better understood by those who have had little experience of the very advanced synthetic chemistry involved, Sol Spiegelman's and Arthur Kornberg's work on living phages may be compared to *the shaping of metals to form valves, camshafts, springs, cogwheels, cylinders, pistons, valve seats and gears in order to allow such components of metal to function, say, as an internal combustion or other engine.*

Such component part shapes formed by *actual information* working on metal *define the nature of the actual and the*

---

[7] A. E. Wilder-Smith, "*The Natural Sciences know Nothing of Evolution*", p. 7 (new edition), TWFT Publishers, Costa Mesa, PO Box 8000, Costa Mesa, CA 92628

[8] A. E. Wilder-Smith, "*The Scientific Alternative to Neo-Darwinian Evolutionary Theory*", pp. 95-103 (new edition), TWFT Publishers, PO Box 8000, Costa Mesa, CA 92628 (where details of the Spiegelman-Kornberg synthesis will be found).

potential information in detail) [9] which is imposed on to the metallic parts by *directionally applied energy* and is *not derivable from any properties which might be intrinsic to the metals themselves – or to randomness even working over immense time spans!* The information required to make the special shapes of the component parts of any machine, including the biological cell, is imposed onto the metals or matter of which biology consists from without *as actual information and not from within the material which takes on the shapes.*

Valve springs and valve shapes suitable for machines never turn up spontaneously to fabricate internal combustion engine component parts no matter how long a time the metal concerned may be exposed to chance or other non–directionally applied influences. Nor do the various shapes of the component parts of biological cells – like enzymes – required to make the biological cell work, ever turn up by spontaneous chemistry even over immense periods of time. This fact is borne out by the observation that the optically active parts responsible for the shapes of certain biological enzymes never do turn up by chance reactions. Specific directional forces must be applied to the component parts for at least a short time if any workable biological or other machine component parts are ever to be arrived at.

Exactly the same consideration applies to the molecular shapes required by the biological machine before it will experimentally function metabolically.

Similarly in the production of the machine component parts of life – (the enzymes for example) the shape of the DNA double helix and other component parts must be directionally

---

[9] "*The Scientific Alternative to Neo Darwinian Evolutionary Theory*", pp. ii-x and Chapter 2

imposed on the matter of the molecule from without. Bits and bytes of *actual information* are required to supply this direction and *not lengthy time spans. Internal non–directional random forces will never do the trick, no matter how great the time spans which are allowed to work on matter.*

The International Symposium held at Jülich in Germany in 1973, which we have mentioned, faces up to the above problems of any and all philosophical theories of a naturalistic materialistic origin of life and evolution and assesses them. But the results reached by the Symposium seem to have passed into oblivion in large areas of the scientific world today – at least as far as biological theories on the origin of life and Darwinian evolution are concerned.

In short, the Jülich International Symposium demonstrated that the laws which govern known chemistry do not allow of the chance synthesis of any optically pure molecule and therefore molecular shapes such as would be vital for the genesis of life from inorganic matter. An example would be the optically pure DNA molecule itself.

The various options which might be adopted to overcome these chemical difficulties associated with chemical shape are examined in detail in the Jülich Symposium. The concept of spontaneous generation is found to be lacking in vital points, so that no scientific basis for any naturalistic origin of life and its vital optical activity could be put forward. It is indeed a great pity that the results of such an international symposium have been largely forgotten especially in biological circles. *For it treats a subject (Darwinian Evolution) which is certainly vital for the philosophy behind pretty well all biological science. The science of biogenesis is surely the fundament on which all biological evolutionary theory rests. If this fundament is unsound, then the whole superstructure of evolution which is built on it is unsound too.*

THE GRAND PROBLEM OF BIOGENESIS TODAY

The grand problem then for biogenesis today is that of attaining *the correct shapes for the component parts of the biological metabolic machine.* For, just as in the case of building an internal combustion engine, *the component parts arise from neither the internal properties of the component parts nor from extended time spans in manufacture, but from the shapes of the components parts which are always imposed on to the matter of those component parts from without the system.*

The component parts of all machines, including those of the biological von Neumann machine, are *always* shaped by directional energy applied to each component part, in fact for relatively short periods of time. This is the standard and most efficient way to make any real machine, including the biological one. This means that *directional actual information is required for each component part for at least a short time.*

What we know today is that such actual information is stored on the DNA molecule, about which Darwin knew precisely nothing. A synthesis between Darwin's outdated views has been sought by some modern information theoreticians by maintaining that such directional actual information arose spontaneously from random directionlessness!

As we have already seen, information renamed as surprise effects has been described by Claude Shannon, but it is useless for shaping molecules to optical purity which are vital for all biogenesis and biological functions. Today it is becoming increasingly clear that this attempted synthesis between the old random theories and modern information theories will not wash.

ORIGIN OF ACTUAL INFORMATION[10]

But there is an additional vital factor which must be considered if a naturalistic origin of life is to be taken seriously in scientific circles. It is the question of the origin of the teleonomic information (*actual information as opposed to potential information)* which characterizes and guides all biological processes.

It is commonly considered that Claude Shannon's Information Theory has solved this problem of the origin of information by demonstrating that the information he describes can be formed by naturalistic stochastic processes alone and might therefore perhaps supply the basis for the biological information observed to be stored on the DNA and other molecules.

The question to be settled at the outset here is that of whether Claude Shannon's information is the same article as the teleonomic actual information with which all biological processes deal. Since this matter is vital for any treatment of the origin of life we have included a chapter on the nature of information according to Claude Shannon and on the synthesis of the teleonomic information with which biology deals and of which it basically consists.

Thus the present treatise includes matters involving the organic chemistry and information theory on which all

---

[10] cf. A.E.Wilder-Smith, "*The Natural Sciences Know Nothing of Evolution*", TWFT Publishers, 1981, and "*Die Naturwissenschaften kennen keine Evolution*", Schwabe & Cie, Basel, Switzerland, 1978, 1980, 1982, 1985. See also A.E.Wilder-Smith, "*The Scientific Alternative to Neo-Darwinian Evolutionary Theory*", 1987, Information Sources and Structures, Chapter 2

biology is built and also a treatment of the nature of information as well as its synthesis, for all biology consists of the most concentrated packet of actual information known to modern science. Whether the dimension known as time could in any way assist us in our study of the origin of life and the origin of actual information is treated too in the chapter on the nature of the time dimension.

Chapter 2
The Nature Of Information

I) The History and Nature of Claude Shannon's Information Theory

It is a relatively well recognized fact today that before any new science can establish itself among already existing sciences it is of primary importance to define exactly the new units of measurement in which the new science intends to operate. Chemistry, for example, had to invent the new units of grams and milligrams before it could develop beyond the stage of the alchemy and the transmutation of base metals like lead into gold by the alchemical methods with which it had worked with no progress throughout the Middle Ages.

Ounces and pounds existed, of course, long before grams and milligrams were established, but before Lavoisier got to work taking the balance seriously in his work, few had used weight as a leading factor in research work in chemistry. Although many alchemists knew that metal calxes (as the metallic oxides were then called) were heavier than the metals from which they were derived.

However, most of the research workers "forgot" to draw the perfectly justifiable conclusion from this observation – namely that during combustion the increase in weight observed in the calx was due to the fact that the oxygen with which the metal combined during combustion to produce the calx showed a positive weight. *Instead they philosophized that the phlogiston which they supposed was emitted during combustion showed the improbable property of having a negative weight!* That is, *on releasing phlogiston during combustion (as they quite wrongly thought), the calx of the metal became heavier, for that which was released – the phlogiston – had a negative weight!*

The old alchemy had worked on the assumption that Priestley's indefinable "fire principle" that is, phlogiston, was released from all metals on combustion, instead of the principle that the metal became heavier in forming the calx because it combined with the oxygen of the air thus becoming heavier than the metal without oxygen. The "fire principle" called phlogiston was a vague term and could not be used as a unit capable of being added or subtracted like the grams and milligrams of the new chemistry could be.

With the balance and the help of the units known as grams and milligrams, Lavoisier and his friends rapidly overtook Priestley ("Dr. Phlogiston") with his diffuse "fire principle" which was not capable of being manipulated scientifically as a unit, like grams or milligrams are and which one could add or subtract at will, that is, be subjected to mathematical manipulation.

When Claude Shannon came to the problem of developing the new science depending on information theory and in which the word processor or computer is at home, he was confronted with a similar problem to that which faced the pioneer French chemists years ago. For although everyone "knew" what information was – just as everyone "knew" what "phlogiston" the "fire principle" was – nobody could actually measure or weigh it!

The grand question was, then, exactly how was one to measure information in *interchangeable* units in developing the new science dependent on information theory? To be of use each unit had to be identical so that addition or subtraction is possible. In order to work scientifically with any information units the first prerequisite was that such units must be absolutely identical amongst themselves, just as milligrams and grams are identical whether they are milligrams of metal in a car or milligrams of muscle in a mammal.

That is, Claude Shannon sought a unit of information which was as interchangeable as grams and milligrams are interchangeable in chemistry and in all branches of science. The sought for unit must not only *qualitatively* describe information but *quantitatively* too.

The great step forward in information theory, in which Claude Shannon showed his genius was in the development of the "binary digit" or "bit" of information as the basic unit of his information theory. Eight "bits" of information constituted one "byte" of information.

But in solving this problem of an identical unit of information Claude Shannon was confronted with serious problems, for one cannot weigh information as one can weigh pieces of muscle. For obviously, when one comes to think of it, one cannot weigh the concepts of which actual information consists. One must remember too that concepts (portions of information) do not consist of identical basic units such as grams and milligrams do.

A practical example will illustrate the point I wish to make here. It would not, for example, be possible to divide up into identical units of information all the information required for building a total steam engine. For the units of information required to build say a valve gear are not identical with those required to build the gear casing. Obviously, there are different kinds of information required for building different parts of various machines – or even of biological cells.

It was in solving this apparently insoluble problem that Claude Shannon showed his genius. How was he to reduce all the various kinds of information required to build steam engines or electro–motors or even biological cells to *identical basic subunits* so as to build up the specific mosaic of total unitary *actual information?* If the unit types of information are different in nature when say an electromotor or even a steam engine, not forgetting the biological cell, is being built, how can one invent one basic unit of information to cover all the various kinds and varieties of information?

Here lay for Claude Shannon the grand conundrum in commencing to work on the new science based on information theory. In solving this problem Shannon went to work by dissecting out of all the various kinds of conceptual actual information one single property. Shannon recognized, of course, that these subunits of information must be absolutely interchangeable and therefore identical – and yet capable of producing the different total information pictures when summated.

Shannon's new dimension of "information" was much like that of the dimension of "time": Everyone thought that he understood just what "time" meant, until such time as he was asked, like Saint Augustine once was, to define it exactly. It was only then – when asked to give an exact definition of what time was, that Augustine realized he could not – and had to resort to prayer for illumination.

An Alternative In Information Theory – The "Black Dots"

Obviously the genetic information required to synthesize a kidney is not identical with that required to build a brain – or a lung.

On the other hand it is clear that identical units of individual black dots such as used in half–toning in printing processes when suitably clustered, can be summated to produce pictures of such *different objects* as an airplane or a motor car or a duck–billed platypus, or even a variety of cogwheels. Black dots in themselves are identical yet their clustering can yield the information necessary to represent such different objects as an airplane, a motor car or even a duck–billed platypus.

Claude Shannon was then, searching for the "black dot" unit of information into which he could analyze all information, no matter how different each object of the actual information might be.

How then could he reduce all the various and different types of conceptual information to one common and basic identical unit – like the "black dots" of the half–toning example?

Shannon succeeded in his search for just such a common "black dot" subunit of information when he chose to look for such elusive subunits of information in the *words and digits used to describe all the various types of information and to transmit them universally.* His logic was that since all the various types of information can be transmitted and handled *in words made up of the common digits of our 26 letter alphabet, he might be able to make a start by using the digits of our alphabet as a basis for his new units of information.*

THE BINARY DIGIT

Thus the "binary digit" or "bit" of information as the basic subunit of information theory was born. It will be clear that Shannon had taken this basic unit of information theory from "logos", that is, from the "word" used universally to describe *all types of information,* as his common unit of information. Thus, being the common basis for the transmission and storage of information and being fundamentally identical in nature, the units could be added and subtracted just like milligrams and grams.

No wonder then that Shannon's Information Theory took off as a new science with phenomenal speed and soon the "word" was conquered world–wide by the computer which is the great machine for word processing.

WORDS, LETTERS, SEQUENCES AND MEANING

It is of paramount importance for understanding the origin of life and of actual information to realize that *the mere formation of sequences of letters of the alphabet to give words – like the adding of the letter "t" to the letters "ca" produces the word "cat",* but that the *sequence of letters in itself does not produce the meaning people who speak English attach to the sequence "cat".* To the Chinese speaking person the sequence "cat" does not conjure up the idea of the feline animal we English speaking people assign to the sequence of letters "cat". *The meaning of any sequence of letters is added afterwards, that is, after its formation, by the language laws and language use one happens to be versed in.*

This fact – that the formation of letters of the alphabet to sequences and words *does not generate any meaning at all – is vital to any understanding of the genetic code and the generation of the meaning and concept of the genetic code.*

Many biological textbook descriptions of the possible mechanisms of the genesis of the actual information on the DNA molecule proceed from the assumption that *the mere summating of the chemical letters which make up the genetic code produce chemical words which gradually but stochastically produce conceptual meaning until the whole sentence thus formed has produced the actual chemical and meaningful word we find in the genetic code today.* The assumption is then made that these stochastically formed words will contain concept and meaning such as we observe today on the DNA molecule.

The words themselves as sequences can in fact be stochastically synthesized, but will contain about as much actual meaning as the stochastically formed sequence "cat" does to a Chinese person who knows no English but who spots the word "cat" in an English text which he does not understand and which carries for him not the slightest weight of meaning.

That stochastically formed sequences contain conceptual actual information and that the conceptual meaning stored on the DNA molecule is thus formed – this error has been copied from edition to edition of biological text books without being corrected for years now. We therefore give here a practical example of how this trick to attempt to explain the origin of concept and meaning stochastically has been played in the text books on unsuspecting students for generations now.

AN EXAMPLE

If we are about to write down some piece of actual information and are going to use our 26 letter alphabet to do so, we have to start the operation with any two letters, – say they were for the sake of argument the letters "ab" we took to start with.

In the English language "ab" does not hold much information, so that "ab" needs in English further letters to make any actual information precise. In languages other than English, as for example in German, where "ab" is in itself quite precise informationally. For "ab" in German means in itself "away from", as in "das Auf und Ab des Lebens" = in English "The ups and downs of life."

If "ab" is to mean anything much in English, other letters have to be added to it first. Thus the next letter or letters added to "ab" will complement its English meaning. This complementation holds the key to the further transmission or the storage of the message. If we add to "ab" the sequence "rupt" then the sequence "abrupt" is produced, which in English contains a very precise piece of actual information.

Shannon measured the addition of a third digit to the first two preceding (binary) ones in terms of a "surprise effect" which is measurable but different in each language used and is valued as a surprise effect according to the identity of the third digit added to "ab." It depends on the *likelihood of the next letter turning up* by chance and being added to "ab". This again is dependent on the language in which one is working and the frequency with which a particular letter turns up in the alphabet from which we are selecting the letter to follow "ab".

If our first two digits had been "aq" instead of "ab", then in the English language *the third letter to be added to "aq" must always be a "u". For in English (but not in certain Arabic based languages) the letter "q" is always, without any exceptions followed by a "u", so that the extra information brought to the digits "aq" by adding the letter "u" is zero in English.* That is the addition of a "u" *has no surprise effect value at all and therefore no extra information is added by this operation.* The effect of adding "u" after the binary digit "aq" is a requirement of the English language and brings no surprise effect and no units of or bits of information with it.

There may be no such rules in other languages so that the bits of information in such languages will be different. If after the formation of "ab" the letter "c" was added then the surprise effect due to added information will be greater than if it had turned out to be "u" after "aq".

Thus it becomes clear that Shannon's bits of information are quite strictly restricted to the degree of surprise effect in adding a further digit to a binary structure of digits. It is most important to realize that the bit of information which Shannon developed was in the interests of mathematical manipulatability, but that it is totally decoupled from concepts or meaning of any sort, that is, it is entirely decoupled from any truly informational meaning or concepts within the ordinary meaning of the word information.

Shannon's bits of information have nothing to do with any hypothetical units of concepts or meaning. *It thus becomes apparent that the genesis of Shannon's bits of information has nothing whatsoever in common with the genesis of thought, actual information or conceptual meaning. Yet it must be clearly kept in mind that the DNA information is in fact thoughtful actual information, holding the results of what may be termed thought on say, how to filter waste metabolic products out of the blood by passing it through a kidney constructed so as to effect this filtration. The functioning of all biological organs betrays basic thought or instructions on how to carry out a biological process.*

The grand confusion in a great deal of mainstream biological research today arises when Shannon's bit and byte units of information, which are units of language conditioned surprise effects and nothing else, are confused with *actual information meaning or instructions. This confusion leads to the mistaken belief that spontaneous stochastic natural chemical processes could lead to the generation of the actual information we find stored on the DNA molecule, when in reality they can only lead to the Shannon type of surprise effect which is devoid of meaning.*

Many researchers even in the Nobel Laureate category still aggressively teach that all biological meaning can be accounted for by spontaneous naturalistic stochastic chemistry. It is now scientifically certain that the *actual information stored on the DNA molecule could never have been produced by any thoughtless stochastic processes. Some scientists have simply not stopped to think the matter out – namely that Shannon's bits of information are basically stochastically produced surprise effects only and have nothing to do with instructions, thought or thought processes, that is with the actual information on which all biology is based.*

The whole misunderstanding has come about by using just one word – namely "information" – in two separate and indeed opposing ways and indeed *without specifically saying so by definition before starting operations.*

Research grants worth millions of dollars annually are being largely wasted in order to finance projects based on *just this misunderstanding and indeed confusion of vocabulary.*

At this point one must remember that the use of alphabetic symbols presents us with the in–built difficulty that the 26 letters of our alphabet *are of course not identical.* The addition, therefore, *of practically any third letter to any system of binary digits must bring with it surprise effects regardless of any actual information involved. The letter "a" is , of course, not identical with the letter "b" which fact obliges us to fabricate a certain "surprise effect" when we add any other different third letter to any binary sequence of letters.* That is, the fact of producing something new as a surprise effect which is vital to Shannon's information theory, is really due to the use of the different letters of our alphabetical system. It has nothing to do with the conceptual meaning or the actual information being processed at all.

If identical "dots" had been used by Shannon as the digital unit, this category of "surprise effect" could have been avoided. For adding an extra dot to two other dots would have produced nothing new which might be mistaken for actual information. What a lot of trouble with materialistic theory would have been saved if this snare had been recognized earlier!

In using the alphabetical system, the use of sequences of non–identical letter sequences instead of the clustering arrangements of identical dots, the whole concept of the surprise effect unit is hidden. However in both systems – whether dots or letters are used – the real actual information lies outside the actual letters or dots used but rather in the spatial arrangements outside of both the letters and the dots.

THE WORK OF PROFESSOR WERNER GITT (GERMANY) ON INFORMATION THEORY

Professor Werner Gitt, who is chairman and Professor at the "Bundesanstalt für Information" (this institution is perhaps one of the most senior institutes of higher learning in Germany dealing with questions of information theory) in Braunschweig, Germany, writes the following: "Der Nachteil dieser Definition des "Bit" ist ebenso offensichtlich: Information nach Shannon unfasst nicht die Information von ihrem Wesensgehalt her, sondern beschränkt sich auf einen ganz besonderen Aspekt, der insbesondere für ihre technische Uebertragung und Speicherung bedeutsam ist. Ob ein Text sinnvoll, verständlich, richtig, falsch, oder ohne Bedeutung ist, wird dabei überhaupt nicht erfasst. Ebenso ausgeklammert bleiben auch die wichtigen Fragen, woher die Information stammt (Sender) und für wen sie bestimmt ist (Empfänger). Es ist für den Shannonschen Informationsbegriff völlig ohne Belang, ob eine Buchstabenreihe einen äusserst bedeutsamen und sinnvollen Text darstellt oder ob sie durch Würfeln zustandegekommen ist. Ja, so paradox es klingt: eine Zufallsfolge von Buchstaben enthält informationstheoretisch betrachtet sogar das Maximum an Informationsgehalt während der entsprechende Wert für einen gleich langen aber sprachlich sinnvollen Text kleiner ist.

Die Shannonsche Informationsdefinition beschränkt sich nur auf einen Aspekt der Information, nämlich dass durch sie etwas Neues ausgedrückt wird. Also nach Shannon ist Informationsgehalt "der Gehalt an Neuem."[11]

(Summary of Translation: Professor Werner Gitt, Chairman and Director of the Federal Institute for Informatics in Braunschweig, Germany writes in the Siemens Journal *63*, No. 4 July/August 1989, pp. 2–7: "The disadvantage of this scheme of things according to Shannon is clearly the following: information according to Shannon does not include the nature of information at all, but restricts itself to a very special aspect of information which is of importance for the technical transmission, storage and retrieval of information. Whether a text possesses any conceptual sense or meaning, whether it is correct or not does not play any role at all within Shannon's meaning of the term information. For whom the

---

[11] Professor Werner Gitt, "*Information die dritte Grundgrösse neben Materie und Energie*", Siemens Zeitschrift, Heft 4, 63 Juli/August 1989 Ss.2-7

text is intended and where it arises plays absolutely no role either within Shannon's meaning of information."

PROFESSOR GITT CONTINUES:

"Indeed, paradoxical though it may appear, it does not matter whether a text is a collection of alphabetical letters arranged entirely by chance or not, from the point of view of Shannon's information theory, a text consisting of alphabetical letters arranged entirely by chance will have a maximum of informational value that is higher than for a text consisting of the same number and identity of letters but which is carefully arranged so as to bear the maximal amount of conceptual information.

For Shannon's definition of information it does not matter whether a text consists of letters of the alphabet chosen entirely at random and without any meaning at all. It may seem to be paradoxical, but a random sequence of letters chosen entirely by chance will, from the point of view of Shannon's information theory, hold more Shannon type of information than a sentence containing the same number of letters but carefully chosen for meaning by a good author."

This means that Shannon has, in fact, decoupled with his term "bit of information" the new unit from any connection at all with our usual expression for concept, information, meaning or teleonomy. That is, what we normally understand by the term information is not coupled in any way with the value of Shannon's bit of information.

The consequence of this fact is, that if one syntheses a bit of Shannon's information by any mechanism, random or not, one has by no means synthesized any meaning or concept at all. That is, no instructions for carrying out a biological synthesis of some organ or other will ever be produced by synthesizing Shannon's type of information. Because the synthesis of all organs, such as kidneys, hearts or brain requires conceptual meaning, i.e. instructions for syntheses. The synthesis of Shannon's bits of information requires only that some surprise effects have been produced, which has no relationship whatsoever to the *synthesis of instructions on, how to, say, wire up a brain.*

One result of this depth of confusion in the relatively new science of information theory has been that many other otherwise experienced research workers in this field have erroneously assumed that the conceptual meaning of any text – even the chemical texts on the DNA molecule – can arise spontaneously without the mediation of the grand thinking organ of the universe – the cerebrum – to supply the necessary thought or instructions behind such concepts and are certainly not related to instructions for the syntheses of biological organs. Shannon's bits of information certainly do arise randomly by sheer chance but instructions such as the DNA instructions for syntheses just as certainly do not. Just as certainly, thought processes do not so arise, although much of the mechanism of thought processes is not yet understood adequately.

Since, however, the information stored on the biological DNA molecule is highly conceptual, instructional and actual, it must be erroneous to believe that such DNA information could have arisen by the same mechanism as Shannon's information can arise. That is, the highly thoughtful instructions and conceptual information required for building say, hearts, kidneys, cerebra or lungs which resides on the DNA molecule cannot have arisen by the chance processes which can lead to the Shannon type of information and which random chemical reactions can certainly produce.

It is at this point that one of the main theories behind modern Chaos theory becomes clear. For the research, say, into the meaning of cloud shapes, which may well in some cases look like human faces, have in fact nothing at all in common with the human visage. That is, these cloud shapes produced by random processes do not carry any associated meaning with the shapes.

It is the confusion surrounding Shannon's term for bits of information with the information which is commonly used to describe concepts, instructions or thought and teleonomy that has led many materialistically minded research workers – even some of them in the Nobel Laureate category – to believe that there is no need to postulate a personal thinking Creator or even Personality behind the Creator.

Often it just does not seem to occur to them that the creator of the instrument for processing thought (the cerebrum) would quite probably be a thinker Himself. That is, that the author of the primate brain, might be a Thinker Himself. Behind the only instrument by means of which thought and concepts are processed (or received) with the help of mere energy in our space–time dimension might a Thinker be reasonably postulated?

For *instructions of most types* are conceived of and formulated by thinkers and persons. It is generally conceded that the information stored on the DNA molecule is in the nature of instructions for building organs or wiring brains. It is thus inconceivable that the information stored on the DNA molecule is of the stochastically formed type.

Might it not be rather less reasonable to assume that an instrument to process thought and instructions was invented and wired by random processes? The view that such a thought processing, instruction dispensing, instrument such as the brain was synthesized in the last analysis by chance processes, this view might to some otherwise reasonable scientists appear to be queer to say the least.

Quite obviously an instrument such as the cerebrum is highly conceptual in nature. Yet some of the foremost intelligences of scientific society have been trying for years to invent just such a machine. That is, one which converts "raw" energy and time into real conceptual teleonomic thought and instructions. That is such workers have been trying for years to develop a computer which really thinks when fueled by "raw" energy and time — like, in fact, the human cerebrum does.

Thus the phenomenon known as thought, instructions or concept has long been erroneously considered capable of arising just as Shannon's bits of information can, that is by random reactions controlled only by naturalistic law and without the aid of an instrument like the biological cerebrum to help it convert "raw" fuel into concept and instructions.

It is the confusion caused by using the term "information" in two entirely contradictory frameworks, but without at the same time specifically and clearly defining exactly either of them, which has given support to the idea that the genetic information stored on the DNA molecule, which is highly conceptual and instructional in nature, could have arisen by random naturalistic processes working over long periods of time.

That is, the incredible confusion in the definition of "information" has given unwarranted support – and a "theoretical" basis – for the idea of a materialistic and indeed a naturalistic origin of biology on earth derived from a primeval chemical soup without any outside supply of actual information to fund the supply of concept and instructions shown in such profusion in all biology. This profusion is seen especially on the DNA molecule which is "bursting" with *coded concepts, that is, instructions for complex chemical syntheses*.

It is precisely the confusion in the terminology of information theory which has given direct credence to the materialistic concepts of life and its origin, seen, for example, in Marxism and Communism, as well as in the left wing often atheistic tendencies in many sections of Western Academe. It is this confusion which has supported and given a pseudo–scientific basis to much militant, political and scientific atheism.

By the same confusion the source of the conceptual origin of life in the thought of the mind of a personal Creator has been displaced and replaced by mere random internal surprise effects in "alphabetical" chemical sequences. Thus, the parameter of meaning or concept in the science of information has become entirely lost in Shannon's information theory. For the origin of teleonomic life is no longer sought in the Mind of a Creator but determined in terms of conceptless random surprise effects alone, rather than in terms of the *basic raison d'être* of instructional information.

The concepts, meaning or sense of information has been lost by the confusion surrounding Shannon's form of reductionism in information theory and it has therefore become respectable in some academic circles to consider that certain teleonomic concepts, such as those stored on the genetic code, can arise by chance chemical processes. It is often believed that concepts or meaning can arise by random processes simply because Shannon's kind of *"informationless" information* (the pure surprise effect masquerading as conceptual information) can and does so arise.

At this point the grand confusion in much of modern mainstream biological origins research on the genetic code and its biological information generation therefore for life itself, has arisen. Even such a celebrity and Nobel Laureate as Manfred Eigen has apparently fallen into this trap, for it seems that he believes that information (without defining exactly what he means by that term) arises: "Der reine Zufall, nichts als der Zufall, die absolute blinde Freiheit als Grundlage des wunderbaren Gebäudes der Evolution..." (J. Monod). (Translation: Pure chance, nothing else but chance, absolute blind freedom, which is the basis of the wonderful edifice of Evolution. J. Monod).[12]

Now if chance really does explain everything biological as J. Monod maintained before he died and which M. Eigen still apparently supports, it must explain too the origin of all the conceptual, meaningful information which is stored on the genetic code. (And also the origin of biological optical activity). The fact would seem to be that some prominent research workers have not adequately analyzed either Shannon's information nor the nature of the actual information which has built and still does build all biology.

Actual and Potential Information – An Example to Demonstrate the Difference Between the Two Phenomena

Shannon's kind of information is really potential and not actual information.

The keyboard of a piano offers us, for example, a store of potential information. Each note is a surprise effect, – for the

---

[12] cf "*Das Spiel*" Piper & Co, München/Zürich 1975, p. 192

constant frequency emitted when struck is, in a certain sense, a surprise effect. *Potentially,* all the music of Mozart or of Beethoven resides in that keyboard.

But it takes the concept–synthesizing biological human cerebrum as well as cerebrum controlled human hands to coax Mozart's "Eine kleine Nachtmusik" that is, *actual information,* out of that keyboard. A cat could run along that keyboard until the cows come home but it would *never* synthesize out of all those *surprise effects of* (the separate notes = the potential information) "Eine kleine Nachtmusik" by its random reaction with each of those notes i.e. bits of potential information no matter how long the cat continued its perambulations on the keyboard.

It takes an active primate type of cerebrum to synthesize "Eine kleine Nachtmusik" from those separate notes (surprise effects, bits of potential information) that is, to coax the actual conceptual information of "Eine kleine Nachtmusik" out of all the potential information on the keyboard.

How illogical it must then be if even senior research workers hope that time and chance will synthesize real conceptual meaning and information out of Shannon's surprise effects. One might just as well hope that chance molecular movement would produce conceptual actual information as that a cat running on the piano keyboard would in time and with perseverance produce a piece of Mozart's immortal music. Or that an empty keyboard would in the course of time peal forth Beethoven's Fifth.

It is a fact of big science that huge and costly research programs are now in progress in certain prestigious universities in European and North American States in which chance and time are supposed to produce the conceptual information necessary for the formation of the conceptual contents and instructions of the genetic code. For some of their scientists believe that all the concepts stored on the biological DNA molecule arose that way and nobody raises an eyebrow – for fear of being classified and ridiculed on "scientific or religious grounds."

Such research is carried out in the attempt to explain on a purely naturalistic basis the origin of the biological conceptual information carried on the DNA molecule. According to the ruling philosophical atmosphere in science today, the origin of such conceptual information is not allowed to be personal but must be inorganic and impersonal if it is to be "scientific" and therefore worthy of a research grant.

A materialistic basis for the origin of life must therefore mean an origin solely within the confines of natural law, which is today erroneously believed to be the only basis upon which any real scientist is allowed to work in origin's research. That is, for life to be explained "scientifically", it must be explained entirely naturalistically – which means without any help from outside of matter and natural law. If materialism cannot be used to explain all problems then the parole is then there is no "scientific" explanation of the problem at all!

FOX AND MILLER'S PHILOSOPHY SUMMED UP

It is at this point that the total inadequacy of Fox and Miller's thesis of biogenesis from racemic amino acids becomes obvious. For by this method only racemates can be synthesized which are useless for the biological synthesis of optically pure enzyme systems and the optically active DNA system to store and retrieve information in three dimensions. The origin of concept and thought by the methods suggested by Fox and Miller remains totally unexplained.

Why are such costly research projects countenanced when it is increasingly clearly becoming obvious that the chances of success are getting progressively slimmer with the passage of time? Might it be that there is an obdurate unwillingness, especially in certain "progressive" academic circles to even countenance the possibility that a Personality, but of course an infinitely greater personality than any human person, – but all the same, a Person with a Will, a plan and a consciousness might be behind all biology and creation? *For behind every machine we know of, a person (or persons) is always the initiant.* Why should biological machines have a different origin in stochastic reactions, which Darwinism directly or indirectly suggests?

The curious fact is that, although no one knows just what the nature of a person is, *all seem to actively and indeed aggressively deny that the Creator could be a person. Why?*

The whole of the Bible, of course, says that all origins lie in the mystery of a Trinitarian Person, who made and conceived it all outside the space and time dimension in an Eternal Mind and that Person became a human person for 30 odd years and taught us His ways of thought and life. Also that this Person wishes for human fellowship, love and cooperation.

But certain scientists appear to hope against hope that there is no such divine Person behind all the obviously conceptual information in biology (but still do not know how to account for conceptual information otherwise than in a person). *Yet the DNA code is just full of the "trademark of persons," that is concept!*

Even animals like birds obviously have a concept of how and where to build their nests. It takes only a look to see that the blackbird building her nest holds a well conceived construction concept both in respect to the nest as also to the siting of the nest. She conceives that leaves will grow on the branch she has chosen and hide effectively the nest in the course of the coming summer. That is, the blackbird has a concept of future leaf growth! But obviously the Creator when He constructed matter did no more with the matter He made than endow it with natural law – but not with the conceptual information as to the form or siting of the nest. Those points are left to the discretion of each individual bird as far as we can see.

The conceptual information required to create a biological person or animal out of inorganic matter had to be added later to matter in an additional act of creation. For it (conceptual information) could not be derived from natural law directly. The conversion could be carried only by the mediation of a cerebrum.

Again other scientists seem to hope, however tenuously, that the material universe itself might be creative, that is, that the material, inorganic universe might produce the side effects of a person, that is produce conceptual actual information or thought – just as the biological cerebrum does.

The idea is this: If chance plus natural law could produce the concepts and conceptual thought which are necessary for biology and for the construction of teleonomic organs like the lung, the heart, the kidney and the cerebrum, then there would be no longer any need to postulate any Divine Personality behind the creation – except perhaps for the origin of natural law stored on inorganic matter itself.

If the inorganic universe itself were creative and therefore productive of thought (as some scientists would like to believe), then have we a reasonable scientific basis for atheism? No, for even if this were the case there would still be need for the Divine Creator postulate – for even then a Creator

– postulate would still be necessary to account for the creativeness of matter itself! In fact we would even then have no basis for real scientific atheism, such as the communists touted for years.

The fact is, that scientifically seen, matter itself does not, – in the lab – seem to be creative. For although it will form geometrically symmetrical crystals due to the valency angles with which the Creator has endowed it, yet it will not, out of this store of in–built information *ever build the concept of any machine spontaneously.* Such behavior, which one might well expect of matter, if it were really creative, has never been experimentally observed. *It always takes a cerebrum to make matter creative – even in the case of the building of birds' nests.*

In science the thought that there might be a thinking personal Creator was an uncomfortable thought and was therefore believed to be extremely undesirable, for such a Creator might have a will of His own and plans we, that is, those who do not want him, would not desire. But if inorganic matter itself were found to be creative, then there is really less need of the postulate of a *personal* Creator at all – with all the possibly undesirable properties of a person!

Even if matter were creative, the origin of that creativity of matter has still to be adequately dealt with. Of course, if the Creator were such an one as Christ was (that is, a person who was more willing to die for his creation than to ask his creation to die for Him), then one might have to revise ones views on the undesirability of a personal Creator – He might well be an exceedingly desirable Person!

That the Creator did not endow matter with conceptual information as a property of matter itself is clearly expressed by the universal validity of the Second Law of

Thermodynamics which states that matter and its structures tend to decay and equilibrium and not to synthesize or to create.

J. Monod believed that chance – that is absolute freedom from constraint of any sort, – synthesized everything including the concepts stored on the DNA molecule. There is no *scientific* evidence for this latter supposition. If the latter point were supported by scientific evidence then Monod was correct in maintaining that this point alone annulled for ever and immediately all religious beliefs because all such religious beliefs would then be without real evidence to support belief in any divine Creator, for chance could then carry out all the "cerebral" thoughtful work of a personal Creator.

Atheism would then be more likely to be true: It would then be more likely that, in view of the unreasonable amount of evil (wars, illness, torture etc.) in the world, it might then seem more reasonable to believe that chance is responsible for everything – or even that some wicked demiurge was responsible for everything!

It is a fact of experiment that, all research carried out up to the present has shown that conceptual actual information does not arise by random processes. *If random processes can produce thoughtful concepts, why did biology go to the extreme trouble of building the most complex of all biological organs – the cerebrum – to produce concept?* Since none of the mandatory optically pure substances required for biogenesis can arise without actual information being present to guide the optical resolution in the absence of an entropy gradient between the two optical isomers – for considerations of entropy gradients cannot here guide the synthesis, but only pattern recognition and concepts – then the Shannon type of information or surprise effect cannot suffice for biogenesis.

There can certainly be no genesis of species without the genesis of life first, even Neo–Darwinism accepts this perfectly logical solution. And biogenesis cannot occur without conceptual thought on account of the fact that the optical activity of the DNA and the optically active enzymes cannot occur experimentally by random processes.

The necessity of actual concepts and actual information is underlined additionally by the necessity of synthesizing not only optically pure isomers for biogenesis but also for the synthesis of the languages and codes of the DNA and other molecules. The fact that the genetic code consists of languages and codes surely points to its origin in personality, *for it is surely only persons who invent and use codes and languages.*

Biogenesis demands the synthesis and separation of not only optical isomers which show among themselves no entropy gradients but only differences of pattern but also the biological languages and codes. These requirements for successful biogenesis surely allow the justified suspicion (at least) that a concept synthesizing language and code using personal Creator was behind the whole project. For as far as we are aware today, *only persons use and produce thought, instructions, languages, codes and concepts.*

Why should then certain of us scientists not feel ourselves justified in suspecting that a kind of cerebrum functioning somewhat like the human cerebrum but on an infinitely larger scale should not have been behind the creation and behind biogenesis at the beginning? How else are we to account for the abundant evidence of conceptual thought everywhere we look in creation and especially in biology, as well as in the structure of matter? *To put all this down to mutation (chance) followed by natural selection is to deny the evidence on hand, since we now know that machine parts and components are always shaped by externally and directionally applied energy and actual information.*

The component parts of no machine extant were ever formed by chance followed by the selection of those that worked best – which is the principle on which natural selection works – for chance forms as we have seen in previous chapters none of the specialized shapes required for metabolic or other machine component parts – those disagreeing with this statement please produce experimental evidence for the chance production of biological or mechanical machine component parts or machines.

Of course it is true that *existing* machine component parts can be *modified by chance.* The point here is that component parts are never formed *de novo* by chance. *Component parts of all machines need actual information to make them in the first place, but then chance can modify them and then natural selection can get to work to select out the ones which work best. But actual information is required in the first place to make the component parts of any machine. Then the functioning machine can obviously and easily be modified by chance. But in order to start the process of evolution by natural selection it is necessary to apply actual conceptual information to inorganic matter first, so as to get biogenesis started. Thus if there is no source of such information no evolution by any natural mechanism can possibly get started.*

This statement is not intended to foster the cause of theistic evolution, because, firstly theistic evolution would not be necessary if there were a Creator present to supply the conceptual information required for biogenesis in the first place. He could easily apply His actual information directly to the species to create new ones without all the waste of time, energy and the cruelty of Darwinian evolutionary processes.

Secondly the mechanisms applied to theistic evolution are wasteful as well as cruel.

I personally do not believe that the Creator, if He was Christ (which is my positive personal conviction) would have crowned His creation by producing man over the dead bodies of billions of innocent lower forms of biology, who through no fault of their own, were eaten or otherwise tortured to death by the stronger and better adapted. If Christ is the Creator, as He claimed to be,[13] then He must have had His tongue in His cheek when He taught that the meek (not the strong) would inherit the earth![14] He was also never wasteful as all evolution, theistic or not, is. Witness the fact that after feeding the multitudes He made the disciples go round and pick up the leftovers.[15]

The Role of the Cerebrum in the Production of Actual Conceptual Information

For conceptual information to arise a special instrument – the primate type of cerebrum – must be called in, otherwise no thought or concepts have ever been known to arise. The primate type of brain is, of course, one of the most complex organs in the known universe and only it has proved its capability of synthesizing (or receiving) thought by converting "raw" energy into thought. *The primate cerebrum is unique in that it thinks.*

The Shannon type of information has, as we have already seen, little in common with thought (information) of the cerebral type. Materialists generally suppose that gradually and for no apparent reason mere surprise effects *are spontaneously* or with the help of certain mystic machines

---

[13] John 10:30

[14] Matt. 5:5, Matt.21:5

[15] Matt. 15:37, Matt. 16:10, Mark 8:8, Mark 8:20

(like hypocycles?) transmuted into thought and conceptual information to supply the needs of matter if it is to become organized up to the living state. J. Monod and others who thought the same way and believed that the intrinsically meaningless sequences of genetic letters produced by random chemistry i.e. surprise effects, gradually became endowed with meaning merely by being exposed to the ravages of eons of time after their formation as mere surprise effects.

There is, of course, a hypothetical machine which we have already mentioned and which is supposed to produce such conversions of meaningless sequences into meaningful ones. It is known in specialized circles as the *hypocycle* and was invented by the well known Nobel Laureate Manfred Eigen, who is active at the University of Göttingen, Germany. In order to effect such a feat as Manfred Eigen claims for his hypocycle it would have to think with the same results as the primate cerebrum does.

However, as far as the literature allows us to judge, the hypocycle does not seem to be wired at all like the cerebrum is wired to perform this feat of producing (or receiving?) thought from mere raw energy, time and surprise effects. Yet it is claimed that it performs the same type of conversions that the cerebrum itself performs. To judge from what has been written about the hypocycle, it is capable of guiding the component inorganic components of life into a functioning metabolic machine.

Our present ability to construct any machine composed of matter is restricted to producing machines which can synthesize almost anything *except genuine thought or concept.* We have up to the present not the vaguest idea just how to set about constructing any concept or thought producing (or receiving) machine. That is, we have at present not the slightest notion of how to set about constructing a true thinking machine, for we do not know at all just how the cerebrum functions – whether it itself manufactures thought or just receives it from elsewhere.

The work of Sir John Eccles[16] with Sir Karl Popper gives a good idea of the complexity of the cerebrum necessary to function in producing (or receiving) thought by using mere energy, surprise effects and time as the raw materials.

The study of biology has shown us, however, that the biologically stored conceptual information on the DNA molecule is capable of supplying the information to build just such thinking machines, for such chemically stored DNA information succeeds in building the primate cerebrum – the most fascinating thinking machine of the universe.

This fact that the information required to wire a thinking machine can be stored chemically provides reason for giving serious scientific thought to the concept that man was created in the image of the God Who calls Himself the Logos, that is, the Word or Thinker, the One who develops thought and expresses it as the Word which conveys and transmits thought (in the form of words). This surely means that man's cerebrum is so wired as to be capable, in a small way, of synthesizing, creating or at least processing the thought behind the creative act.

THE POSSIBILITY OF A UNIVERSAL THINK–TANK?

The reader will have noticed that whenever we have mentioned the origin of thought we have usually added the possibility that the thought content was not synthesized in the cerebrum but that it might have been received from a source

---

16 "*The Self and its Brain*", Springer Verlag International, 1981, Heidelberg, London, New York and Berlin

elsewhere. There is a school of thought which maintains that the cerebrum is merely an instrument for receiving thought from a sort of universal think–tank. Such a think–tank might act as a reservoir for all the thought in the universal dimensions and that the suitably conditioned brain could under some circumstances eavesdrop on this universal think tank.

Such a circumstance might be useful in explaining some of the prophetic gifts which prophets like the biblical Daniel undoubtedly possessed. Eccles seems to think, judging by some of his remarks, that the human cerebrum is capable of looking outside time into other dimensions outside time. The Apostle Paul taught the same concept, when he spoke of having been "outside the body" and having experienced matters which were unlawful for a mortal human to have seen.[17]

One would have thought that more intensive research work would have been done and more grant money spent on research designed specifically to elicit from the structure of the chemically stored information on the genetic molecule just how the information required to build a thought producing or receiving structure like the human cerebrum is precisely built.

As far as I know, science today has no clue on how to convert meaningless digital sequences into meaningful ones. That is, just how the cerebrum thinks. "Hypocycles" are apparently not even modeled on the machine (the cerebrum) which we know can effect the feat of synthesizing (or receiving) thought.

What is really needed is work on the structure of the cerebrum so as to throw light on how the cerebrum comes to process its thought.

Surely it would be useful to apply one's energy to working out just how the genetic code supplies matter with the information to wire a thought processing machine like the human cerebrum does. For, if one machine (the cerebrum) can pull off the feat of coming by thought why should we assume that no man–made machine could imitate the biological machine in being capable of real thought?

However, considering man's present track record in recent and in long past history, it might turn out to be a terrible disaster for all mankind if ever man should succeed in manufacturing a real thinking machine. He might use it to boost his own capability of thinking and increase thereby his own capacity for producing more deviltries than even those he

---

[17] II Cor. 12:2-4

has produced already! *For increasing his intelligence does not mean at the same time increasing his good will.* If the will is not improved at the same time as man's intellectual capacity, then increasing his purely intellectual capacity ought to be fought shy of. For he has done badly enough being in charge of the restricted amount of such intellect as he has!

Chapter 3
The Space–Time Continuum: A General Introduction

It was Albert Einstein who taught scientists and others to think in terms of dimensions when thinking of time, so that today one thinks less of time than of space–time. Let us first of all try to clarify the concept of this double dimension known as space–time. Most theories of biogenesis teach today that huge quantities of the time dimension are needed to convert mere surprise effects into the concepts and thought of biology.

Thus we need now to examine the time dimension itself to assess whether the synthetic properties alleged to reside in the time dimension are really to be found there. Is the nature of the time dimension such that it alone could simulate the cerebral type of creativity needed to synthesize or receive thought and concepts such as we see them in such profusion in biology?

Perhaps the simplest way to conceptualize the idea of space–time is by considering the concept of a world–line first.

This will familiarize us with the notion of another double dimension first.[18] (See figure 5 p. 69)

First of all, what exactly is a world–line? A concrete example of what a world–line is will be the most helpful way to proceed: Consider the flight of an airplane from one city to another. Its position before, during the flight and afterwards may be seen from an inspection of the graph as shown in Fig. 5, which includes the parking before and after landing.

Time AB is spent in parking at Washington airport. Time BC is spent flying between Washington and New York. Time CD is spent parked in New York airport.

The world–line of the aircraft is given by the line ABCD. That is, the trajectory of the world–line represents the movement of the aircraft in respect to space and time, that is, its motion in the space–time continuum.

---

18 A. E. Wilder-Smith, "*Creation of Life*", pp. 153-156, TWFT Publishers, P.O. Box 8000, Costa Mesa, CA 92628, U.S.A., and Pro Universitate e.V., Roggern, CH - 3646 Einigen, Switzerland. Originally published under the title of "*Erschaffung des Lebens*" by the Hänssler Verlag, Neuhausen/Stuttgart, Germany

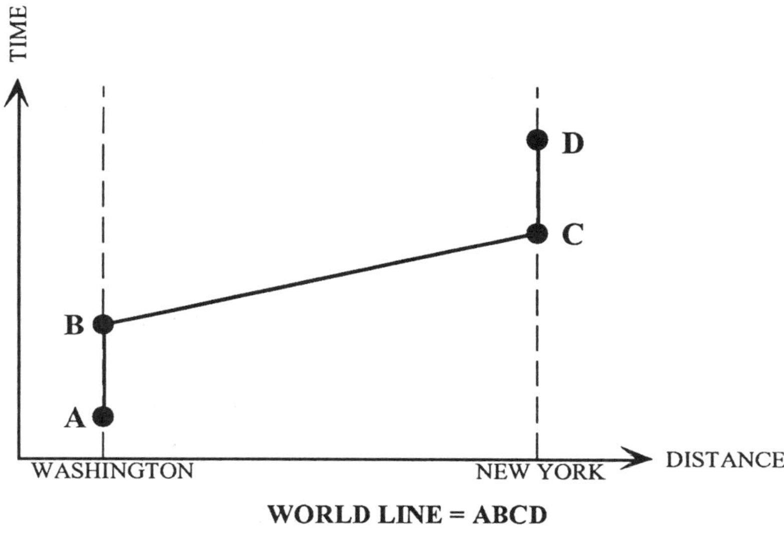

WORLD LINE = ABCD

Figure 5

B) THE NATURE OF THE SPACE–TIME DIMENSION

In a series of lectures given in 1985 in Oxford England on the "Nature of Time" and published by Basil Blackwell, Oxford and New York, edited by Raymond Flood and Michael Lockwood, some 400 auditors turned out to hear some of the quite abstruse lectures on this subject.

The interest in the subject of "The Nature of Time" is very great indeed, in spite of the fact that quite advanced physics is often needed in the elucidation of the subject. "Lay people" did not seem to be put off attending the lectures by the high standard of scientific understanding required to keep pace with the treatment of the subject. Surprisingly, many "lay people" are sufficiently interested in the subject of the "Nature of Time" to take on quite advanced physics in the pursuit of the understanding of the "Nature of Time".

We suggest here that the apparent inability of modern thought to accept the idea of a short period of time (that is the biblical six days of creation) as sufficient to create the world and all that exists in it[19] is perhaps largely due to a lack of understanding of the nature of the space–time continuum. Similarly a lack of comprehension of just what *creatio ex nihilo* (creation out of nothing) as far as the creation of time itself is concerned, means, can sometimes be cured by going into dimension theory in greater detail.

The difficulties of accepting a creation in six active creative days can be minimized once it is realized that, according to the Bible and its understanding of dimension theory, *man and biology were not derived from the dimension of time and space at all, (that is from space and time) but rather from a synthesis completed in the eternal Mind of the Creator where the dimension of time does not rule at all.* That is, biology and man were created in timeless eternity and then rapidly placed after conception in the mind of God, into the time dimension.

That is, eternal thought, conceived of and worked out in the eternal dimensions (timelessness), made us. What, however, if the already completed work of Eternal Thought were transferred in a short time (that is six days) to space and time dimensions? Today there are no scientific difficulties about such a concept, for even in time and space vast quantities of conceptual thought and actual information which are conceived of in Europe, are daily transmitted across the Atlantic at the speed of light. The *thought–work done in Europe during months* can very easily be transmitted in the shortest space of time to the other side of the Atlantic by fax through telephone wires or radio at the speed of light.

---

[19] Genesis 1

The invention of the fax machine has made this rapid transmission of already completed thought practicable. If all biological synthesis was, in fact, completed in eternity – *if that feat is feasible, then the transmission of the results of that eternal thought synthesis rapidly into space–time is scientifically really no problem at all.*

Thus, the six days of creation should offer few difficulties to the scientist who believes in God and His report on just how creation took place in six informational surges in six days from eternity into time, according to the Genesis report. If creation took place in the eternal Mind and was completed in concept in Eternity there would be little difficulty from a scientific viewpoint in the transmission of the completed eternal thought concept (that is the Creation) from eternity into time in six signal surges, each of one day's (our time) duration. Each signal surge would be complete and perfect on reaching our time dimension.

It may take 20 years to work out the complete design of a Rolls Royce in the drawing shop. But in a suitably equipped workshop that complete concept may be executed in matter in a day or two or transmitted by FAX machine at the speed of light in seconds of our time.

MULTIDIMENSIONAL REALITY

Until comparatively recently some physicists were inclined to believe that the space–time continuum (dimension) in which we live, made up the whole of reality. That is, our time dimension is the totality of reality. Such is, in effect, the materialist position. That is that, "the whole shooting match", the total extent of reality, was summed up in our space–time continuum.

As Professor Davies in his book "God and the New Physics" points out, many leading physicists have today abandoned the materialist position and believe today that there are many other dimensions or realities besides our own space–time continuum. Some physicists even believe that there are up to eleven other dimensions outside the space–time dimension. Thus, the "here and now" of our space–time dimension, in which some materialists believe even today represents the whole of reality. The whole "shooting match" of our space–time continuum is certainly not the whole of reality. Materialists often used to mock at certain "gullible people" who believed in other dimensions such as heaven or hell. This is no longer the case today amongst the unknowledgeable – at least.

The fact that some otherwise non–religious physicists have come to believe in up to eleven dimensions outside our own space–time continuum speaks a very clear language, one leading our thought away from the old type of materialism that could only believe in what it could touch and see. Other dimensions such as those of heaven and its opposite pole are no longer the field of activity of the metaphysicians only but also of the physicists and the psychologists.

Celebrities such as Sir Karl Popper also believe in World I, World II and World III, and spend a great deal of thought and time on elaborating the contents of these various dimensions of concept.

New Dimensions in Space, Time and Space–Time

Isaac Barrow, one of Isaac Newton's mentors, remarked that because mathematics frequently makes use of the term "time", mathematicians ought to have a quite distinct idea of the meaning of the word, otherwise they are quacks. "My auditors," he said, "may therefore on this occasion very justly require an answer of me, which I shall now give and that in the plainest and least ambiguous expressions, avoiding as much as possible all trifling and empty words."[20]

In the first place we should carefully note that the term "time" is an abstract one. We cannot point to it such as we can to a dog or to a house and say "that is it". For like Hume's soul, the moment (of time) itself entirely eludes our grasp. Even if we look for help in a dictionary, we find that "time is continued duration." If we then look up "duration" we are enlightened to learn that "duration is a part of time". So that definitions of time of this order do not help us much in our quest for enlightenment.

Augustine once said that he knew just what time was until someone asked him for a precise definition. Then he knew no longer but had to resort to prayer for enlightenment.

VIEWS ON THE NATURE OF TIME (LEIBNITZ)

Leibnitz postulated a reductionist view of the nature of time which necessitated distinguishing between events taking place in time from temporal items such as moments and instants. Similarly in the case of space, there is the need to distinguish between things in space such as bodies and spatial items such as locations.

Leibnitz's approach to the question of time is strictly reductionist, for he thought that time is nothing over and above an ordered system of events and that similarly space is nothing over and above an ordered system of bodies in space.

Leibnitz's reductionist view maybe compared to the statement that the average person has 2.2 children. If one then proceeds on the basis of this statement to search for a person

---

[20] W. H. Newton Smith, Space, Time and Space-Time, in "*The Nature of Time*", p. 23, Editors Raymond Flood and Michael Lockwood, Basil Blackwell Publishers, Oxford and New York, 1986

with 2.2 children one is going to be disappointed, for there exists in reality no such person with 2.2 children. Such a person is mysterious and indeed mythical. He arises from the mistake of specifying "the average reader" to be on a par with "the youngest reader." Here again there is no such thing as "the average reader" over and above "the reader of this paper".

Leibnitz believed that such statements as "the beginning of time" really require translation before they can be taken at face value. Such a translation can be best effected by noting that they refer to events in time and to the temporal relationships between events in time. Such statements say nothing as to the nature of time itself but concern only events taking place in time.

In an exactly parallel manner Leibnitz viewed space. Space itself, he said, is nothing. Space consists only of the bodies in it which make anything out of it. It is merely the spatial relationships between bodies in space which give space any significance at all. Talk of space is then only a way of dealing with the relationships of bodies in space.

That is, in Leibnitz's view, talk about time and space is merely to be construed as talk about events and objects in time and space. Whether time had a beginning or not is to be settled by asking whether there was ever a first event in the history of the universe. The value of the term "space" depends in this view, entirely on the contents of space, that is, on the bodies occupying that space. Similarly time itself possesses no meaning beyond that of events taking place in it.

Views On The Nature Of Time (Sir Isaac Newton)

In contrast to the above views of Leibnitz, there are, of course, other views on the nature of Time. Sir Isaac Newton, for example, was convinced that *both space and time have an existence of their own which is entirely independent of their contents*. It was the Neoplatonists who taught this view to Newton.

A technical consequence was that, if God had not created the physical world, space and time would none the less have existed, even if only as receptacles into which He could then have placed His creation, that is the *events* and *bodies* which He created. Thus space and time in this view have their own existence apart from the bodies and events which might have been laid into them at the Creation.

In these philosophies Newton and Leibnitz could not agree. Both of them diverged from the orthodox Christian teaching on these subjects, which is that space and time and all that in them is (heaven and earth) were all created by the Creator, "in the beginning".

THE NATURE OF THOUGHT AND COMPUTERS: CAN COMPUTERS THINK?

It is surely as unscientific to teach that matter and space are creative, that is to say capable of thought, as it is to believe that time, left to itself, is creative.

There has existed in the scientific literature a long standing feud among certain experts as to whether the computer can really think or not. We do know for certain that the computer can calculate at a rate that no human cerebrum can hope to match. But the question is, is calculating, as such, the same thing as thinking? For the thinking mathematician not only knows what he is doing when he calculates, whereas the computer when it is calculating certainly does not know or understand what it is doing. The machine is certainly not thinking when it calculates unless it understands what it is doing while it calculates, which, of course, no machine yet invented does.

As far as we know, no man–made machine has ever brought forth conceptual thought. No machine has ever shown any indication of being conscious of its own calculations. It does not know that it is calculating. *But machines and computers can all produce surprise effects and yet produce no thought at all.* The idea that the "ravages of time" can convert surprise effects into thought is merely another way of maintaining that time (and perhaps space too) can think – that is be creative in their own right. The idea is quite as bad as maintaining that inorganic matter can think and spawn concepts.

We urgently need to find out exactly what mechanism the primate brain uses to think (or alternatively how it manages to receive conceptual thought). For thought might be connected with the article or person known as logos outside time and space and not be a product of time and space at all, which would mean that the cerebrum would be an instrument for connecting transcendental thought or logos with the dimension of time and space.

This is just a possibility to bear in mind when dealing with such abstruse subjects as consciousness and concept. Once we know how the primate brain manages to think or to receive thought as the case may be, once we know that, it is on the cards that man may be able to invent a similar instrument himself, capable of doing likewise. An instrument which, with the help of time and energy might become self–aware so as to produce consciousness and conceptual thought. For the two properties probably belong together. But we are definitely a long way from achieving such aims at present.

Perhaps it is a crowning mercy of God's providence that man has not yet been able to invent a really intelligent thinking, conscious machine. Even with his present dumb machines which are certainly not capable of conscious thought at all, he has done enough damage to God's creation and to man's own self as well as to his world that perhaps he could not be trusted with conscious machines. Perhaps we ought not to be given the chance of inventing intelligent thoughtful machines to supplement our present awful track record.

The subject of meaning and thought together with consciousness has become in recent years almost irrelevant as far as Shannon's information theory is concerned. Yet the nature of both consciousness and thought are both burning questions in the study of today's physics, psychology and biogenesis (the origin of the DNA molecule with its three dimensional information storage and retrieval system). Most things are held by modern materialists (including J. Monod, of course) to be reproducible in terms of conceptless random surprise effects. Presumably, this would include the nature of consciousness and self–awareness too.

JACQUES MONOD'S MATERIALISM

J. Monod said that *everything* could be so explained. As far as I know he did not exclude the nature of consciousness or concept from this statement, although no one knows what either are! To maintain that randomness could have made both consciousness and thought without even knowing what either are, was surely a rather rash statement of doctrine to have made?

Thought consists essentially of envisaging the synthesis of whole material or other hierarchies in representative form by conscious or unconscious simulation in the mind, maybe before realizing the same conceptual hierarchies later in material form. The computer is capable of this feat, although it may not be conscious of the process. Simulation in actual material form with the material building blocks in matter represents the materialization of thought.

That is, whole machines can be built in building blocks of thought rather than in those of blocks of actual matter for the actual machine as opposed to the "machine in thought". Conscious thought is thus "creating in the realm of the imagination," conceiving in the mind, that which it is intended to construct in hard matter maybe later.

THE PSYCHOSPACE OF THE MIND AND ITS CONDITIONING

It is possible of course to construct not only machines in the mind, but also pictures of landscapes, of animals and of other human beings as well. All these subjects can be projected at will in the mind, which is full of all sorts of projections which can be summoned up into the psychospace at will. Such pictures make up the thought content of the mind. *Thus the mind can be thought of as a type of space, in fact, a psychospace. Just as space is occupied by material bodies, so psychospace is occupied by thought objects.*

It is every bit as important to fill the space of our space–time continuum with good and useful objects in it as it is important to fill the psychospace of the mind with useful and wholesome thought–objects. For just as the bodies in space determine the nature of space, so the kind of thoughts we load into our mind daily condition the whole of our thought world. When one considers the programs of "chaff" which are unloaded onto the unsuspecting populace daily by means of television and other media one can hardly be surprised at the signs of decay rampant in our society.

When one considers the enormous costs of television this corruption of the minds of the people with "chaff" is found to be not only ruining the people financially – and somewhat accounting for the huge national debt afflicting the nation – but the corruption has gone so far that a president can get elected by solemnly and officially promising not to raise taxes and yet at the same time deal with the national debt problem! Yet practically the first official act that the same president initiates after having been elected on this platform of not raising taxes is the raising of taxes!

So degenerate has the psychospace of the people become that the most powerful man in the world can get elected on an untruth without the threat of impeachment. Free elections can use untruth with impunity for the most powerful political office in the free world.

THE SUBSTANCE OF THOUGHT

Other objects not made of matter can be used for construction purposes by thought. It is, of course, possible to project arguments and theorems in the same manner in the psychospace of the mind. Thought is, so to say, simulative construction, constructing, but not with material blocks but with building blocks made of thought material in the psychospace.

The thinker goes through in thought all the steps required for putting material building blocks together, but does it in the first place in the psychospace before going on to try to do the same in arranging material blocks in space–time. The building up of theorems and arguments gives practice in the conditioning of the whole psychospace and thus building personality and character.

If the thought has been realistic it will have taken into account all the possible snags which could turn up in the space–time construction project, correcting such in the psychospace first to prevent them (the snags) turning up in space–time later.

The Nature of the Psychospace

It is said that Wolfgang Amadeus Mozart[21] could call up in his mind complete scores of music and project them there for accurate correction and testing in his psychospace without hearing the music actually played on any instrument in space–time.

Some of his scores took hours to play through on actual material instruments. In his mind Mozart could actually envisage whole scores and correct them there. This could take place on his own testimony "in a twinkling of the eye", though the actual score might have taken hours to play through in space–time. Mozart's psychospace seemed to work in a *different time frame to the actual time frame required by the physical music* in space–time.

One can therefore begin to understand that our minds (or psychospace) were never constructed to be filled with "chaff" of the kind it is fed daily to the populace by some types of television (or other) programs. For the psychospace was conceived of in eternal dimensions and is satisfied therefore with nothing less valuable than real eternal matters.

Here again we find that Sir John Eccles' surmise that the cerebrum can work in other time frames to be not so far off the truth as some materialists might claim.

The Time Dimension is Not Creative

As far as the laws of thermodynamics are concerned the adding of time to any system hierarchy or order reduces the system, hierarchy or order to increased disorder. This fact has been observed in all scientific experiments to date under controlled conditions and has proved to be so universal that it

---

21 Roger Penrose, "*The Emperor's New Mind*", p. 575, Vintage, London, Melbourne, Sydney, Auckland, Johannesburg, 1989

has been summarized as such in the Second Law of Thermodynamics. (see Fig. 6 p. 81)

THE THREE LAWS OF THERMODYNAMICS

1st Law:
Matter can neither be created nor destroyed.

2nd Law:
The total energy of the cosmos remains constant but the amount of energy available for useful work is always diminishing.

3rd Law:
As absolute zero temperature (-273°C) is approached in a perfect crystal, it's entropy will also approach zero. *One can never attain absolute zero temperature.*

Figure 6

The thesis that life with all its optical activity and information storage and retrieval systems demonstrates such an enormous reduction of entropy and increase in conceptual information that its arising *in toto* in chance chemical processes would, on the surface of things, seem to fly in the face of the Second Law.

One can, of course, overcome the consequences of the Second Law by the use of machinery. A self–winding watch does this trick by using random wrist movements to wind up the main spring *that is to decrease the entropy of the machine by chance.* The important point is that such *self–winding mechanisms never arise by chance* themselves, they need conceptual actual information to be imposed on to the metal of the watch. And that actual information necessary to reduce entropy by chance never arises by random processes itself.

In a parallel manner life needed conceptual information to get it started, randomness could not achieve that feat alone without conceptual information to initiate its necessary machinery. This all adds up to the observed facts that life and its conceptual information never arise by naturalistic random mechanisms. *The case for the creation (rather than the evolution) of the super machinery of biology is thereby made.* For all machines arise by applying information from without matter to matter – rather than by evolution out of the properties of matter.

Darwin and the early evolutionists had, of course, not the slightest notion of the machine and chemical complexity of even the ameba, let alone of the human cerebrum. For this reason the idea that the ameba arose spontaneously out of a primeval slime deserves to be reckoned with mythology and wishful thinking.

Louis Pasteur finally and scientifically dispelled the idea of spontaneous generation over a hundred years ago now. But it dies, this myth, very slowly, for it is a myth humankind apparently would like to believe to be true, for it suits his acting as if there were no Divine Creator throughout his life.

Even natural selection could never separate higher forms of life from lower ones unless advanced chemistry could have produced at least first of all a lower form of life to be operated on by natural selection. But we know of no naturalistic chemistry which could achieve this feat of spontaneously producing even the lower forms of life.

It is therefore a mistake to conceive of natural selection in time as creative, for natural selection can only separate out what is already present in biology and can synthesize nothing biological *de novo*. Chemistry guided by conceptual information has to do that.

Chapter 4
The Theoretical And Experimental Basis Of Geological And Evolutionary Dating

Creation and Measuring the Time Dimension

One finds on looking through popular magazines and geological text books that the age of the universe and the earth is conceived of in millions if not billions of years. Although perhaps a majority of conservative scientists believe that the dimensions of time and matter are both created dimensions the logical conclusion of this conviction is scarcely ever followed, namely that, if the time dimension is an *ex nihilo* created dimension, then before that dimension came to be created, there could be no time, *that is no age to measure. Our minds are built to work mainly in the time dimension* and cannot function if there is no time in which to work.

Among the rather complex consequences of the Evolution / Creation debate of recent years *is the emergence of the fact that few students seem to be able to grasp the huge consequences of the fact that a real creative act would have on the dating methods used by science.* To determine the age of the universe and the time spans available for the so called evolution of life from any primeval slime, radiometric methods (some of which are listed below) are used.

To expose the theoretical and practical difficulties in dating methods which would occur if a *creatio ex nihilo* (creation from nothing) has ever taken place in past history, the following examination question was once set in a general knowledge test at one of the senior state universities in the Middle West.

The examination question was couched in the following terms: *The radiometric dating of radioactive ores.* Examiners had often noticed in interviewing students that they had often swallowed whole aspects of radioactive dating methods but had never thought them through maturely. Therefore it was thought advisable to put some simple questions in this area in the general knowledge tests which all students had to undergo. As coming academically trained persons it was thought that an examination question on this general subject would induce more mature thought on the subject.

The examiner retired to the laboratory balance room an hour or so before the examination hall was opened and weighed out there a mixture containing 50 mgs. of radium as a sulfate salt (that is the weight of radium metal was in toto 50 mgs.) and 50 mgs. of metallic lead also as the sulfate salt.

The mixture was then labeled as a crude ore and the examinees were told that the given ore contained 50 mgs. of radium and 50 mgs. of inactive lead. The examinees were also told that the half-life of radium was for the purposes of this examination 50 thousand years.

The examination question to be answered was: *What is the age of the ore in thousands of years? Date the ore and state your reasons.*

Almost all the students turned in their papers giving as the age of the ore 50 thousand years, their reasoning being that 50 thousand years was the given half-life of the radium on its decomposition route to inactive lead. Since exactly 50 % of the radium had "obviously" decomposed to inactive lead, then the half-life of the radium would show that exactly half of the radioactive radium had already decomposed to lead, so that the age of the sample was merely a case of simple mathematics. The ore was 50 thousand years old.

The examinees who gave that answer – and that was most of them – were each given a passing grade only. The correct answer should have been, of course: the information given us is insufficient to reliably date the ore at all. For the vital concentrations of both the radium and the lead at *the beginning of the experiment were not stated and without them a reliable dating of the ore is not possible.*

Without this precise information on the *starting concentrations* no absolute dating of the ore is theoretically possible. In actual fact, of course, we privately know that the sample of "ore" was made up by the examiner a few hours before the examination – in the balance room. A "creation" of the "ore" had occurred in that balance room – in the mind of the examiner – and it was only by *excluding* this possibility that the students answered without considering this creation possibility in their mind and fell into the trap.

Had they thought to ask what the starting concentrations of radium and lead were, the correct answer would have been in sight. In field tests of radioactive ores the same questions ought to be asked if a really reliable dating answer is to be gained. In field test dating, of course, the scientist can only *guess* at the starting concentrations of the radium and the lead. If the scientist resorts to the "simple mathematics answer" he is, in his mind, perhaps unconsciously cutting out the possibility of a "creation in the balance room."

The moral is that other errors too are fallen into by the practice of excluding in the mind the practical possibility of a *creatio ex nihilo* in the past. Experimental error even in science can occur unless we take *creatio ex nihilo* very seriously. There is so much other evidence for a creative act at the foundation of all nature, that our mind set should take it seriously if similar errors as the above reported one are to be avoided. The greatest danger lies in the fact that the 50 thousand years half-life was much too facile to be really true.

This brings us back once more to the sample given to the students in the Middle West State University to date on the basis only of the concentration of radioactive and inactive material present in it. The students "forgot" that if a synthesis, that is, if a "creation" had taken place in the balance room a few hours before the examination started, *there was no absolute way of finding that out from radiation and radiation end products by radiometric dating.* During any synthesis such as creation would represent, the normal processes of decay are simply not taking place and these processes are the very means by which we measure time in radiometric dating processes. Thus the prerequisites of any dating calculations i.e. constant rates for decay processes are in abeyance in any synthetic (that is creative) act.

If creative processes have taken place during any part of the world's history then they must annul and cancel out the very decay processes on which dating in general relies.

Long Term Dating Methods

How is, then, long term dating carried out? An example is better than long philosophical explanations where the volume of words tends to hide the real meaning of those many words:

Radioactive elements such as radium or thorium decompose to other radioactive or inactive elements through a long chain of radioactive stages which are well known. Each stage decomposes to the next stage at a set, predetermined rate so that if one knows experimentally the concentration of each decomposition stage present, *then one ought to be able to work out mathematically how long it took a particular sample to reach any particular stage.*

Thus various isotopes of lead are reached at the end stages of the uranium decay chains. The lead isotope concentrations then will give an idea of how long it took for a given uranium sample to reach a particular stage of decay.

But there are various hitches to the above perfectly good theoretical scheme for determining the age of a radioactive sample. This fact can be best explained by taking the Potassium/Argon method first. Potassium is radioactive and decomposes radioactively to the noble gas known as argon. The decomposition rate to argon from potassium has been determined with considerable accuracy. Thus, theoretically, it is a case of "just catching the amount of argon given off from a given weight of potassium and the age determination requires just a simple calculation."

A certain amount of potassium will yield a certain amount of argon in a certain amount of time. As the rate of argon production is pretty well independent of the physical conditions surrounding the sample, this radiometric determination would seem to be plain sailing.

But the practice of the dating experiment is vastly different. For argon is, as pointed out above, a noble gas, that is, it does not combine chemically with other substances. The argon on being released from potassium escapes into the crystal lattice of the potassium salt. Here it does not combine chemically with anything that might fix it firmly but becomes physically only *loosely entangled in the crystal lattices from which it is easily disentangled.* For example, heat and pressure changes can easily displace the argon from its mere physical entanglement in potassium salt crystals. Then the sample of argon is lost to the outside atmosphere. *If all the argon has been lost from the crystal then the sample shows according to this method no age, for it has apparently not had the time necessary to accumulate a good large sample of argon.*

On the other hand since there is argon in the atmosphere surrounding the sample, some argon from the air can easily diffuse back into the sample. By which back–diffusion into the crystal the experimenter is deceived into believing that it actually came from the potassium present. *This particular sample contaminated with atmospheric argon must then appear older than it really is.*

EXPERIMENTS WITH LONG TERM DATING CALIBRATION IN ICELAND: THE CALIBRATION OF THE POTASSIUM/ARGON DATING METHOD

The volcanic island of Surtsey off the coast of Iceland appeared a few decades ago suddenly out of the sea. It came up in an enormous cloud of steam as molten lava from the ocean floor streamed upwards towards to the surface of the ocean. It was colonized by plant and bird life as soon as it had cooled.

A scientist wished to calibrate the potassium argon method by determining the age of the island of Surtsey by the Potassium/Argon method.[22] *Thus the age of Surtsey was found by the Potassium / Argon method to be many millions of years old whereas by actual history it was known to be only a few tens of years old.*

---

22 See "*New Scientist*", July 3rd. 1975 p. 20

What exactly had happened? Argon from the atmosphere had diffused back into the cooling lava and given it thus a false age.

First of all the argon from the potassium decay had been totally driven off from the potassium lattices by the heat of the molten lava. Then the now cold lava absorbed argon from the atmosphere which thus produced the apparent great age of the young, now cold lava laden with argon absorbed from the air – but not generated from decaying potassium.

NO UNCALIBRATED LONG TERM RADIOMETRIC DATING METHOD TO BE TAKEN SERIOUSLY

The lesson to be learned from this calibrating experiment is simple. It is that:

No long term dating method has any scientific value whatsoever unless it has been adequately calibrated on samples of certainly known age. This is the case for all scientific methods – and not only for dating calibration. Calibration is the iron rule which all serious scientists rigidly observe. We can only ask ourselves if certain "old earthers" have taken this rigid scientific rule sufficiently to heart when they solemnly announce that the earth is 4.55 billions of years old? The amount of argon in certain potassium ores "proves" this "scientific fact"!

It is therefore in the eyes of real science little more than mythology when certain astronomers seriously announce in the name of science that the Genesis report on creation is in serious *scientific* difficulty! In actual fact it is the "old earthers" and astronomers who are in real scientific difficulty for not having learned a primary lesson in science which I had to learn *before* I started work on my first Ph.D.

My professor took me into his laboratory to carry out my first experiments as a graduate student. I thought he would assign me to experiments in the use of some complex costly instrument or other. But instead he set me to work on *calibrating the old instruments in every corner of the lab.* I felt really humbled, for I was much too self–important to take on such humdrum work.

My professor was a canny Scotsman and insisted on calibration *done by myself so that I would learn to critically assess the meaning of real reliability and the experimental capability of all my instruments thoroughly.* He "knew his way around in the laboratory" and insisted on my learning "first things first." "Old earthers" and even some astronomers sometimes just do not question the reliability of their methods or their instruments. For they often seem to have forgotten the importance of the rigid rule of all real science:

"Calibrate your instruments and methods first before you launch into wonderful mathematics but using faulty uncalibrated data. They have often not learned to believe *nothing, no results at all,* from uncalibrated instruments.

The rule of real scientists is to accept no data, indeed *nothing at all from instruments that have not been calibrated against known and certain norms.*

The results of their "scientific mythology" are endless disputes especially among Bible believing Christians. Scientists ought to desist publishing anything on these "results" at all – to say nothing of not even starting their calculations until they have learned that *the kingpin of all serious science lies in taking no notice at all of any data, until calibration against known age samples and norms has been thoroughly and exhaustively carried out.*

In principle, because the same thing that has happened with the potassium/argon method can happen even with the lead isotope method (lead isotopes can diffuse into the system or leave the system and nobody knows which process, if any, has occurred).

Therefore, it is *absolutely necessary to calibrate all these methods against a known aged sample.* Because, however, one cannot always do this in view of the large ages with which the Potassium/Argon and the Uranium/Lead Isotope methods have to do, all ages determined by these methods before calibration against a sample of certainly known age are more than highly suspect: they are indeed *absolutely worthless* unless calibration against samples of known age has been carried out thoroughly. Of course in principle it is just such a calibration that cannot be done – there are few known samples of sufficient and certain age to use as a norm. [23]

Of course, evolutionary theory itself has been used as the norm, but that is the equivalent of lifting oneself out of the bog by pulling on one's own boot straps. It is using the theory to prove the theory.

Even the lead isotope method is subject to suspicion, for there are few samples of certainly known age to calibrate it against. It is not that an isotope of lead is volatile like argon, but that lead can enter or leave the system, thus falsifying the age determination. Unless a method or a sample can be calibrated against a sample of known and certain age, no method of long term dating is absolutely fool–proof or certain.

---

[23] See the Surtsey example cited.

THE C^{14} METHOD OF DATING

Because the C^{14} method works in thousands of years rather than in millions, it was successfully calibrated against tree rings up to nearly 10,000 years old (actually up to round about 6000 years using the bristle cone pines in California). But when it comes to the potassium argon and the lead isotope methods *there are few samples of certainly known age* against which one can calibrate the millions of years which might come into question here. There is one notable exception to this calibration difficulty which we have already cited in the now classical case of the Island of Surtsey.

THE LEAD ISOTOPE METHOD OF DATING

Let us look again, this time a little more closely, at the lead isotope method of dating on which the latest estimated age of the earth (4.55 billion years) rests.[24]

According to Faure the age of 4.55 billion years was established by Patterson in 1956 who had analyzed three stone and two iron meteorites for lead isotopes and compared them with one sample of oceanic sediment. The results all fell on one straight line so that Patterson concluded that the age of the meteorites was the same as that of the earth, that is allegedly 4.55 billion years old.

Faure maintained that he had found that "deep sea sediment contains lead whose isotopic composition varies regionally and not all of them fit the meteorite isochron as well as the sample which Patterson analyzed." Faure, instead of updating Patterson's original date, adopted it unchanged, *knowing it to be incorrect.*

Alexander Williams asks himself why Faure did this. Gale, Arden and Hutchinson obtained more data on meteorite lead isotope ratios and *using the same reasoning as Patterson they came to the conclusion that the earth had a negative age!*

Would it not be safer and more scientific to say that until we *can calibrate all our isotope methods against samples of known age (which is practically the universal practice in other branches of scientific research) it would be more scientific to say that we have no idea of the real age of the earth or the*

---

24 Alexander R. Williams, "*Long Age Isotope Dating Short on Credibility*", Creation ex nihilo, Technical Journal, Vol. 6, Part 1, 1992 pages 2-5, ex nihilo, Creation Research Foundation, Sunnybank, Brisbane, Queensland, Australia

universe? Except, of course, that which was revealed to us from the One Who created it all?

For further information on Long Age Isotope dating it is recommended that the paper by Alexander R. Williams in "Technical Journal" be "studied, marked, learned and inwardly digested". To reject biblical time tables and dating on the basis of experiments carried out using uncalibrated instruments and methods would be termed in some strictly scientific research laboratories "pseudo scientific."

The Surtsey (Iceland) dating by the Potassium / Argon method, constitutes an excellent calibration method, for the age of the Island of Surtsey was historically well known. This sure historical method showed that Potassium Argon was totally unreliable when materials of known age were found, on which the method could be reliably calibrated.

Chapter V
Some Recent Fossil Evidence

1) FRESH HADROSAUR FOSSILS IN ALASKA[25]

In the region of the Colville River in Alaska extensive beds of *fresh Hadrosaur bones* have been found in recent years, which fact caused a good deal of heart searching in some geological circles. For these beds of bones of large animals were found to be so fresh (recent) that they were considered to be bison bones by some people.

Samples of these fresh bones were then sent to experts further South who quickly established that they were Hadrosaur bones, that is duck billed vegetarian dinosaurs. The bones were run across by accident during the search for oil by the oil companies working on the North Slope area of Alaska where so much oil has been found in recent years.

The author of the above cited paper considered that the bones must be Cretaceous that is 50 or more millions of years old to fit in with other geological Darwinian evolutionary and radiometric dating of dinosaur bones. But some experienced chemists do not believe that fresh bones from which intact proteins have been recovered could retain traces of undecomposed proteins for millions of years. Bones that old – many millions of years old – would not contain any

---

25 Kyle L. Davies, "*Journal of Paleontology*", vol. *51*, No. 1, 1987, pp. 198-200

undecomposed proteins, for the proteins are not that stable chemically.

Gerard Muyzer of Leiden University in the Netherlands and his colleagues report using immunological tests to identify a specific bone protein (osteocalcin) in several dinosaur fossils that (allegedly) date back 75 to 150 million years. [26] "If it (the fresh protein) is indigenous" says Lisa Robbins, a micro paleontologist of the University of South Florida in Tampa, "then it is the oldest protein".

The headaches in the area of dating are nearly always caused by taking at face value the millions of years yielded by uncalibrated radiometric instruments and Darwinian evolutionary theory. Surely it would be much better to take seriously for dating purposes the *known instability* of proteins and the dating information yielded by these instability readings on proteins rather than speculative radiometric uncalibrated data?

It must be remembered that uncalibrated and therefore on principle unreliable radiometric dating methods are habitually used to date the fossil geological formations which are the foundation stone of modern Darwinian theory. In plain language this means that the very basis of a large proportion of modern biological Darwinian science is based on unscientific mythology at least as far as experimental dating science is concerned.

The above mentioned research workers are convinced that the osteocalcin they found in the bones did not arise from bacteria or invertebrates (by contamination) for neither of these organisms produces osteocalcin. So the protein must be genuine dinosaur protein and not derived from invertebrates or

---

26 "*Geology*", October 1992

bacteria. *That means plainly that the dinosaur proteins from dinosaur bones are in fact relatively young.*

Other researchers are justifiably skeptical about the genuineness of the dinosaur protein found because they cannot believe that any proteins would last that long under normal conditions. *Rather than question the dating methods on which all evolutionary theory depends, the findings of chemists as to the stability of proteins are ignored in the interests of sustaining the long time periods necessary to sustain evolutionary theory.*

Evolutionary theory has been maintained all these years by simply making it and its dating of the geological layers the absolute criterion for all dating used. The question of calibrating accurately the dating methods before launching into speculative evolutionary theory is seldom even raised. There was little to check the dates given by evolutionary theory against. Evolutionary theory *constructed* the dates on the basis of evolutionary theory.

The dates were based on speculative evolutionary theory and these dates not surprisingly confirmed evolutionary theory! That appeared to be sufficient for evolutionists of a certain category! No wonder that students and others could never find anything radically wrong with the whole theory! It has been self–sustaining and self–confirming for years now. Nobody dares to question the genuineness of the dating of the dinosaurs, not even when fresh protein was recovered from their bones, for no one dares to suggest that the dates have always been unverifiable. *That is, in Carl Poppers terms, the dates have been "unfalsifiable" and therefore utterly unscientific.*

Just a little exactness and practical scientific know–how in examining the shaky results yielded by long term radiometric dating would have straightened out these problems of the chemical instability of animal proteins long ago. The incompatibility with known chemical facts (instability) with the alleged age of the dinosaurs at their extinction has gone uninvestigated for far too long now.

Muyzer and his colleagues (Leiden University) had hoped to isolate the osteocalcin from Dinosaur bones and then determine its amino acid sequences. By comparing the dinosaur sequences with those of the osteocalcin of birds and crocodiles they hoped to fix the evolutionary relationships of dinosaurs, birds and crocodiles.

All this kind of speculative work is done in preference to doing the really necessary work, namely that of *examining critically the dating methods on which all evolutionary theory is firmly based.*[27]

David D. Gillette describes a new large Sauropoda Diplodocidae recently discovered 16 Km west of San Ysidro, Sandoval County, New Mexico U.S.A. [28] Approximately 25 vertebrae (caudals and sacrals), a partial pelvis, several ribs, five chevrons and approximately 230 gastroliths have been recovered from the new large sauropod. The extraordinary size of the skeletal elements indicates that this individual may be the longest dinosaur known. In life it approached the mass of the largest sauropod dinosaurs. On account of its record–breaking size the new find was named Seismosaurus halli. Its axial length was between 39 and 52 meters.

---

[27] *"Science News",* October 3rd. 1992, p. 213, "*Protein identified in Dinosaur Fossils*"

[28] "*The Journal of Vertebrate Paleontology*", II(4): pp. 417- 433, December 1991

Of course, new fossil finds are by no means restricted to the animal world, for the plant world too has yielded new material of breathtaking proportions. The site of the new finds in the plant world is the Arctic region of Northern Canada. The new light thrown by these finds concerns dating problems too, a sample of which we will now describe briefly:

2) Fresh Dawn Redwood Forest found in Northern Canada. [29]

The scientific journal "Canadian Geographic" describes a fossil forest of massive tree stumps in the Axel Heiberg Island, Northern Canada in August 1985. Paul Tudge, assigned to the Geological Survey of Canada, was piloting his helicopter over the Axel Heiberg Island in August 1985 when he noticed some unusual objects on a barren hillside. He landed to investigate and saw that the whole hill was littered with wood and tree stumps. James F. Basinger then made plans for the 1986 season.

Basinger published some most impressive photographs of the findings he and his team made at the Axel Heiberg Island site – a site which is being kept secret for the present to avoid the risk of vandalism, if the site becomes too well known. One might be tempted to suspect that other grounds for keeping the site secret might be valid too, grounds such as will become clear when we have described the nature of the findings at the site concerned.

---

[29] "*Our 'Tropical' Arctic, Forest of Massive Stumps discovered in our polar region, reveals the vastly different world of 45 millions years ago*". Text and photos by James F. Basinger, Canadian Geographic, 1986-1987, Band 106, No. 6, pp. 28-37

We intend to cite some of the findings verbatim so as to allow the reader to assess objectively for himself the nature of these findings and then to make up his mind as to the age of the forest found so near to the geographical North Pole:

"Walking among the ancient stumps and logs it is so easy to let the imagination erase tens of millions of years (!A.E.W-S), to step not over fossil, but over freshly fallen trunks, almost to feel the lushness of the forest. It is a feeling I have never before experienced, for nowhere can the contrast between the ancient and modern environments be so marked as that between the richly forested swamps of millions of years ago and the barren polar desert of Axel Heiberg Island today."

"This sense of being transported into the past is made all the more real by the remarkable preservation of the fossils. *The wood has not been altered through all this time. It looks and feels almost like freshly cut* wood – it splits and splinters, it can be carved with a knife and it burns as readily as kindling. Most of all it still has that reddish hue we often find in softwood lumber. The only thing missing is the scent, the oils having evaporated long ago."

"Forty–five million years old! yet we are picking up pieces of wood, digging silt away from stumps still rooted in ancient soil, and lifting thick mats of conifer leaves from near the bases of trees. What had preserved these fossils for so long? Why had they not been lost to time and decay?"

"Far rarer though were catastrophic events that could preserve an entire forest. *Such events could only have been floods of immense proportions, carrying huge quantities of sediment into the river systems and spilling out over the flood plain, rapidly burying the lowland swamp forests beneath a suffocating blanket of silt. Such events occurring once in perhaps tens of thousands of years would envelop entire forests preserving roots, stumps, logs and even the litter and soil of the forest floor. Dead spires protruding through the sediment would soon rot and topple over as new forests colonized the area."*

Comment by the present author: The last five words "new forests colonized the area" reveal a whole world of make–believe, for there is not a shred of evidence in the whole valuable paper by Basinger that *new forests ever re–colonized the area of the original forest. If they ever had done so, they would have thoroughly destroyed the forest which has now been found.* The *immense flood of which he speaks was obviously followed by a rapid and radical change of climate and temperature, a change so radical that no forest ever replaced the primeval one buried by the huge flood which Basinger postulates.*

Here we have, I believe, some concrete evidence for Noah's Flood which, too, was followed by a radical change of world climate such that no new forest ever re–colonized the area of the original fossil forest. For immediately after the Flood the Scripture assures us that the seasons (summer and winter, seed time and harvest) started.[30] Before the Flood there was a mild climate world wide, with no polar ice caps. The growth of the redwood forest so near to the geographical North Pole proves that point.

Basinger continues: "Another destructive process, coalification, has been avoided at the fossil site. Coalification – the metamorphism of woody debris to coal – requires very deep burial, where the temperature is high, for many millions of years. Our fossil forest was never buried more than a few

---

[30] Genesis 8:22

hundred meters and has rested in a relatively cool part of the earth. In fact for the last few million years it has probably been frozen solid. This has been of utmost importance for preservation of the mats of leafy litter on the forest floor. These leafy mats are the upper undecayed layers of forest soils. Even moderate decomposition would have reduced the litter to a spongy humus. Again their fine state of preservation attests to rapid burial beneath a protective blanket of silt. Had coalification proceeded to any extent at all, the mats would have been reduced to an amorphous mass. But the leafy branches are easily separated and lifted away."

COMMENT ON BASINGER'S ABOVE STATEMENTS:

In contrast to what Basinger asserts, it is not a fact that for coalification to occur, deep burial *for millions of years* is required, for quite recent experiments have shown that the telephone poles running across the hot deserts in Southern Australia, have yielded at their bases a first class quality of coal in a few short years at Australian normal temperatures *without deep burial to provide pressure. Laboratory experiments have demonstrated repeatedly that wood may be converted into first class coal in a matter of minutes provided the temperature and pressure are optimal.*

Basinger then continues: "*Perhaps the most important aspect of the forest is not that it will tell us what* grew in the far North 45 million years ago, but *how* it grew. The high latitudes then enjoyed a much milder climate than at present due to circulation of warm ocean currents through the Arctic Ocean and perhaps higher levels of carbon dioxide in the atmosphere, an early version of the "greenhouse effect" we hear about today. The Bering Strait may have been somewhat wider (then) to allow warm Pacific water to enter the polar region and the greenhouse effect may have occurred a number of times before in the distant past."

More important than even temperature is of course the fact that redwood forests must have sunlight. At the present site there is now some 6 months of relative darkness, which fact alone, apart from temperature considerations, would forbid the growth of such a redwood forest. Thus these facts would pretty well prove that the whole catastrophe of the entombed redwood forest at Axel Heiberg was probably caused by a change in the angle at which planet earth was inclined towards the sun to some 23 degrees out of the perpendicular.

It would seem to the present author to be highly unlikely that the *Bering Strait would open up quickly enough to supply the warm water for climate changes*. No sign of new colonization by secondary forests on the site of the now extant fossil forest ever developed. Whatever change of climate occurred at the site of the forest took place so quickly that after the present fossil forest had died of suffocation by silt, not a trace of new forest developed. Six months of relative darkness would take care that no more redwoods ever grew again on the site regardless of the temperature and warm Pacific water. Basinger described no signs of secondary growth after the primeval forest had died. That is, the change of climate and *maybe lighting* must have been very sudden indeed.

It would seem to be much more likely that some sudden change in the axis of the earth towards the sun took place. If the earth had encountered a large asteroid, that is, one of a sufficient mass to wobble the planet earth on its axis of rotation, during the wobbling the axis might well have been modified to its present some 23 degrees out of the perpendicular, which would cause seasonal weather to appear immediately afterwards, together with huge tidal waves which would swill water around the earth as apparently happened during and after the Flood.

All this would cause a sudden change in climate accompanied by a huge flood and the institution of seasonal weather just as the biblical report suggests, together with arctic ice caps. The changes in the weight of water on the ocean floors might precipitate tectonic activity too accompanied by the resultant pushing up of high mountains in place of the low profiled hills of the pre–Flood world with its *universal* mild climatic conditions.

But in respect of the flood theory, the new finds on Axel Heiberg Island certainly do *not* tell us *how* the climatic changes occurred (Basinger), but they certainly do tell us that it was the *result of a huge flood followed by a sudden and radical change of climate and probably lighting conditions. That is the plain simple message of the Axel Heiberg Island site.* Perhaps these facts contributed to the decision to keep the site secret!

In fact the history of the newly found forest on Axel Heiberg Island would seem to be strangely congruent with the biblical report on the subject of the Deluge described by Noah. The remarkable freshness of the dinosaur bones and the dawn Redwood trees found on Axel Heiberg Island confirm the relatively recent date of the same Flood. For the hadrosaurs were vegetarian and needed luxuriant tropical plant life, and therefore much light, to live on! Perhaps they died out when the forests perished. Thus the theory that an asteroid type of catastrophe contributed to their demise may contain some truth – for that might have altered the lighting of the arctic regions.

One must remember too that the older geologists who propounded the huge age of the earth and the universe did not have the modern Arctic fossil animal and plant findings at their disposal. It is therefore time to begin revising our views.

For the older views were obviously based on sketchy fossil evidence. The new finds put the whole evolutionary concept behind biology into a vastly different light today.

The sketchy geology known at the time of Darwin may have justified some of his views then. But modern finds compel us to *thoroughly* rethink all the rather primitive ideas which Darwin and his friends developed over one hundred years ago. Before the nature of the DNA molecule was even guessed at, one could have accepted perhaps a great deal of what Darwin said. But today new finds, as always, demand often entirely new ways of thought. Of course, it will be most difficult for any radical change in views to take place, for many older professors have built up their careers in Darwinian ideas. Such will most reluctantly accept any newer, radically opposed ideas to those of Darwin.

SOME OF THE COSTS OF THE DARWINIAN THEORY TO THE TAXPAYER

We have mentioned in the foregoing text that the Darwinian error brings with it large costs in wasted time and money. We now shall cite some of these costs concretely in projects and cost estimates:

Since Darwin's evolutionary theory is counted in all mainline science today as established fact – few doubt that primitive life was generated in the first place by naturalistic law which was allowed to work over huge time spans and that it evolved upwards entirely by natural selection to the higher animals including homo sapiens sapiens – who has as a result of "evolutionary fact" produced a technically trained society boasting of electronic communications which are indeed formidable.

Therefore, the modern argument now runs: If such a high state of technical civilization has developed from inorganic matter as a result of Darwinian based evolution on earth why should parallel development to a technically developed civilization not have occurred under suitable conditions elsewhere than on the earth? Say in the outer planetary systems, which we *believe* to exist in outer space? For such systems, it is argued, are built of the same kind of matter as the earth. The same kind of time dimension can work on this matter out there as it allegedly has done in space–time on earth.

Is there, therefore, not a chance at least that intelligence such as ours has developed spontaneously in outer space where and if the conditions under which matter exists were suitable? Is it not on the cards that intelligence such as ours has developed from matter in the course of billions of years just as we have allegedly evolved on earth? Therefore, say the Darwinians, let us look at outer space with a view to finding out if such advanced civilizations have evolved as they allegedly did here on earth.

For maybe they too "out there" have developed radiotelephony just as we have and are trying to contact us, especially is this the case if these societies "out there" have discovered (like we have!) the laws of Darwinism, for then they allegedly will "know" that wherever matter and time exist together under favorable conditions there the laws of Darwinism will determine the evolution of life and intelligence, even "out there."

Wherever the existence of matter under the right conditions of time and temperature is a fact of science, there the possibility of intelligent beings and cultures produced by Darwinian "laws" of evolution must be considered possible. If those civilizations "out there" have indeed discovered Darwin's evolutionary laws then they will "know" that other such civilizations will possibly exist where matter and time exist and will have probably developed upwards to intelligence just as we allegedly have here on earth. It is therefore on the cards that the presumed outer space intellects will be trying to contact other civilizations elsewhere in other planets too.

A search has therefore been set up by NASA named "Search for Extra Terrestrial Intelligence". On October 12th 1991 the Anniversary of Christopher Columbus' landing in America, SETI (= Search for Extra Terrestrial Intelligence) was set up by NASA to look for the realization of Darwin's ideas in outer space.

How is SETI to look for the signs of Extra Terrestrial Intelligence? The answer is, "the same way as Professor Lovell did some years ago." Professor Lovell trained his radiotelescope in England on to a very dense pulsar, which rotated so quickly that it emitted rapid radio pulses which Lovell at first mistook for Morse code like pulses and thought to be intelligent coded messages. But further computer work showed that there were no codes or messages hidden in the pulses at all. But in the early stages of the work, the pulses were thought to be coded messages and the imagined source of them was nicknamed "Little Green Men" (= L.G.M.)

Thus the whole idea of SETI is founded on the allegedly *universal* truth of Darwinian Law, which is supposed to be a proven fact of science today although its base in Spontaneous Generation was disproved years ago by the great Louis Pasteur. Now Darwinian enthusiasts are to be allowed to spend some 100 Million dollars over ten years in this search for intelligence in outer space. Darwin's biggest hoax of all time in World Academe is one of the most expensive jokes played on Academe in recent years.

Why do Scientists allow it to go on when the evidence against it is so clean cut to anyone who has the time and knowledge to expose it? The reason is perfectly clear: *Because a scientist's career in most sciences including biology is effectively mined if he does anything much to expose Darwinian Evolution.* An additional factor is, of course, that science is taught so that the known facts of science which contradict what Darwin said are very tenuously taught and so taught that students do not know the bearing, say, of the facts of optical asymmetry on spontaneous generation and biogenesis. Nor do they know that Shannon's information has nothing to do with what is commonly understood as information.

If intelligent life is present "out there" It will have been originated by the combination of actual information with matter just as Arthur Kornberg and Sol Spiegelman demonstrated when they synthesized a living phage from inorganic matter by adding actual information to suitable matter.

ADDENDUM

In the foregoing text we have spoken a good deal of optical activity, identical entropy status and mirror images which chemistry, unless supplied with actual information, cannot produce.

Before life could get started at biogenesis, chemistry must have been assisted by suitable actual information to overcome the fact that the lack of an entropy gradient between optical isomers prevents it from providing the close proximity and fit needed to start metabolism and to provide energy for biological metabolism at biogenesis.

If, now, the supply of substances from unguided chemistry consists of a mixture of left handed and right handed isomers for building say biogenetical proteins, the resulting sequences of amino acids in the protein will result is a total "higgle–de–piggledy" of left handed and right handed molecules (see Fig. 1 p. 18) which will not slot into the receptor sites on the cell. The result will be that no energy will be released to finance the cell metabolism.

MIRROR IMAGES AND ENTROPY STATUS

Ordinarily, if alpha amino acids are produced by mere organic chemical means say from ammonia, methane and water, alanine and other amino acids will be produced according to the scheme developed by Fox and Miller. But whatever amino acids are produced each will consist of equal quantities of mirror images, which stand in the same relation to one another as I do to my image in my mirror. (See Fig. 1 p. 18).

Now it is obvious that a mirror image of the person standing in front of the mirror is of the same constitution and composition (if the mirror image were present in real form and not merely as a reflection in the looking glass) as I myself am. It is obvious too that the degree of structure and complexity of the mirror image is identical to the complexity of me myself. That is, in fact, just what it means when we say that the *entropy status* of myself and my mirror image is identical.

In other words, the measure of chaos needed to be overcome to synthesize myself from nothing is identical for myself as it is for the mirror image of myself. It is just as difficult to make me from nothing as it would be to make my mirror image from nothing – if the mirror image were made of real matter and not merely a shadow or a reflection.

To put this concept in banal language it would be just as hard to make my mirror image in matter as it is to make me in matter – the entropy status is identical for the mirror image as it is for me. The sole difference between the two lies in the fact that the mirror image is made of left handed isomers whereas the original of the mirror image is made as it were of the opposite optical isomers, say in this case the right handed isomers. And to sort out which isomers chemistry ought to make, chemistry needs actual information – because here it cannot work on entropy differences as it normally does.

ENZYMES, THE BASIS OF THE BIOLOGICAL PROCUREMENT OF METABOLIC ENERGY: THE NECESSITY OF BIOLOGICAL "FIT"

The food on which all biology depends for life is to be compared to a "safe" in which valuables are stored. The "valuables" which are stored in food are represented by the energy which is present in all biological food and which the cell must have at its disposition if life is to be lived by the cell. For without energy the cell can neither reproduce nor divide. Neither can the cell in higher forms of life develop any nervous activity or movement without having energy at its disposition. This is the first requirement of all biological life.

Now, there are two methods of opening the "safe" known as food to get at the energy hidden in it:

1) One can break open the safe with the help of an acetylene torch and maybe a crowbar. But this high temperature and violent method is a rather an inelegant method of getting at the energy of food.

2) A more elegant method of getting at the contents of any safe (including that of obtaining the energy in food) would consist in finding the correct key to the safe and then turning it *gently*. If done correctly the key will operate the tumblers in the key–lock system, thus opening the door of the safe to let one get at the valuables (the energy, that is) hidden in the food.

ENZYME – SUBSTRATE SYSTEM

ENZYME	**SUBSTRATE**
("Key" or "Hand")	("Lock" or "Glove")
Esterase	Ester
Pectase	Pectin
Lipase	Glyceride
Cholinesterase	Acetylcholine
Lactase	Lactose
Rennin	Casein
Pepsin	Protein
Trypsin	Protein
Urease	Urea
Arginase	Arginine
Catalase	Hydrogen Peroxide

Figure 7

It must be remembered here that if even just one of the many profiles on the key we have chosen to open the safe with, is incorrect, the key will jam, the door will not open and the energy in the safe will not become available. This is a pretty exact model of the principle on which all biological enzyme systems work and on which the metabolism of all life metabolism depends. The enzyme is the key and the acceptor site is the substrate on which the enzyme works. Figure 7 (p. 113) gives a list of enzyme systems together with their "locks" into which they fit to give the close proximity and fit needed to allow the system to work and release the energy of the food, the energy of which is needed for cell metabolism.

Just as in the case of the gold or other valuables in the safe the safe can be opened by employing one of two methods: one can employ the acetylene torch method plus crowbar to the safe and so release the valuables in the safe. One could set the whole frying pan full of bacon and eggs (i.e. food) on fire and so release the calories therein contained in a smoky reddish flame. But the method is rather looked down upon in the circle of good chefs.

They prefer to release the energy in the bacon and eggs by encouraging the ingestion of the savory bacon and eggs into the mouth and stomach where the same work of releasing the hidden energy can be done elegantly and enzymatically at low temperature in the gastric juices.

This method is far more elegant and satisfactory than that of setting the frying pan on fire. The enzymes are far more elegant in obtaining a really close proximity by the "hand and glove fit method" such as optical activity and perfect fit allows.

But this method requires a lot of prior work to get the chemistry guided into the correct mirror image system so as to provide the correct isomers which give the correct key fit, thus avoiding the open flame method featuring a burning frying pan.

Figure 7 (p. 113) shows the system of keys and locks found ubiquitously in nature which is capable of opening various "safes" and extracting their energy. If the "safe" or food containing the energy happens to be an ester, then the type of key that will fit it will be an "esterase" key that will open that kind of "safe".

One of the remarkable facts of chemistry is, that if the energy contained in any "safe" which can be termed an ester is required then an "esterase" is the "key" which can open it to release its energy for the metabolism of the cell.

These lock and key mechanisms are known in scientific circles as enzyme and substrate systems. (See Fig. 6 p. 81). They must fit into each other very exactly in order to function. The "fit" is in these cases dependent on optical activity as we have already discussed.

In figure 4 page 18, I have shown diagrammatically a chain of left handed gloves fitting into a chain of left hands coupled to one another by thumb and little finger. It will be obvious that if one has in a chain of left hands even just one right hand that one fact will destroy the "fit" of the whole chain, because it destroys the proximity necessary for all enzymatic activity.

These facts of chemistry are well known but scarcely ever applied in the theoretical considerations of Darwinism and the spontaneous biogenesis on which all Darwinian theory depends. These facts arise presumably because Darwin himself knew just nothing of such technicalities. For if in a chain of optically active amino acids constituting an enzyme just one of the chain members is substituted by a wrong handed member, then the biological enzymatic activity will destroyed by this change, for the "fit" is totally destroyed by just this one change in profile. And naturalistic chemistry is not capable of supplying the optical purity mandatory for supplying enzymatic fit.

It should be obvious to any thoughtful person that the universal occurrence of keys and locks which fit one another everywhere can be interpreted as evidence of universal planning and universal forethought. Surely it is inconceivable that complex chemical locks and keys, hands and gloves, that fit one another perfectly and universally are the products of Darwinian chance stochastic mechanisms. Any mechanism at all implies thought and plan.

Index

actual information, 32, 41, 42, 43, 44, 45, 50, 52, 56, 58, 59, 60, 70, 82, 110, 112
actual matter, 78
age determination, 89
age, 93, 94, 95, 100
Alaska, 97
alchemists, 35
alchemy, 36
alphabet, 39, 40, 41, 44, 47
amoeba, 1, 22, 82
amino acid, 5, 6, 9, 23, 54, 100, 111, 115
ammonia, 5, 111
Andrews, 20
Apostle Paul, 64
Arabic, 42
Arctic, 101, 104, 106
Arden, 95
argon, 90, 91
asteroid, 105
Atheism, 58
Augustine, 73
Axel Heiberg Island, 101, 106
axis of the earth, 105
back–diffusion, 90
basic subunit of information theory, 40
Basinger, 103, 104, 106
Basinger, James F., 101
Beethoven, 53
Bering Strait, 104, 105
Bible, 55,70, 92
biblical six days of creation, 70
binary digit, 40
binary sequence, 45
biogenesis, 2, 8, 9, 10, 16, 17, 19, 21, 24, 27, 30, 31, 54, 58, 59, 60, 67, 77, 110
biological cell, 6,38
biological cerebrum, 56
biological conceptual information, 54
biological enzyme systems, 113
biological life, 112
biological machine, 29
biological metabolic machine, 31
biological person, 56
biological process, 43
biological synthesis, 47, 71
biological teleonomic information, 10
birds, 55, 57, 100
bison bones, 97
bit of information, 40, 43, 49
black dots, 39
blackbird, 55
Blackwell,Basil, 69
blood, 43
brain, 39, 48, 49, 61
bristle cone pines, 94
Bundesanstalt für Information, 45
byte, 37
C14 method, 94
calibration, 92, 93
California, 94
calx, 35, 36
Canadian Geographic, 101
carbon atom, 15
carbon dioxide, 104
cat, 53
cell, 11, 112
cerebral, 58
cerebrum, 48, 49, 50, 53, 56, 57, 58, 59, 61, 62, 63, 65, 76, 80
Chaos theory, 49

chaos, 111
chemical reactions, 16
chemically stored information, 65
Chinese, 40
Christ, 57, 61
Christian, 75, 92
Columbus, Christopher , 109
cloud shapes, 49
coalification, 103, 104
coded concepts, 51
codes, 59
Colville River, 97
Communism, 51
computer, 36, 40, 50, 75, 77
concept, 49, 50, 55, 62, 67, 76
conceptual information, 47, 48, 53, 55, 56, 58, 60, 61, 62, 63, 81, 82, 83
conceptual thought, 50, 56, 59, 70, 76
conifer leaves, 102
creatio ex nihilo, 88
creation, 20, 59, 71, 75, 82, 85, 87, 88, 91
creative, 83, 88
Creator, 2, 8, 19, 20, 49, 51, 55, 56, 57, 58, 60, 61, 75
Cretaceous, 97
crocodiles, 100
crude ore, 86
crystal lattice, 89
Daniel, 64
Darwin, 3, 31, 82, 107, 109, 115
Darwinian Evolution, 19, 26, 30, 110
Darwinian evolutionary processes, 60
Darwinian evolutionary theory, 98
Darwinian theory, 115
Darwinian thesis, 20
Darwinian, 1, 10, 22, 24, 97, 98
Darwinism, 55
date, 87, 88
dating calculations, 88
dating calibration, 91
dating methods, 85, 100
dating, 87, 88, 89, 95, 98, 99
Davies, 71
Dawkins, Richard, 21
Dawn Redwood, 101
decay, 88, 89, 91
decaying potassium, 91
decomposition rate, 89
Deluge, 106
digit, 39, 42
dimension theory, 70
dimensions, 67
  of time, 85
dinosaur, 97, 98, 99, 100
directional actual information, 31
Divine
  Creator, 56, 82
  Person, 55, 56
DNA, 8, 10, 19, 23, 24, 25, 26, 29, 30, 31, 41, 43, 48, 49, 50, 54, 55, 58, 59, 63, 77, 107
dots, 45
double helix, 23, 25, 29
duck–billed platypus, 39
duration, 73
earth, 75, 95, 108
Eccles, Sir John , 63, 64, 80
Eigen, Manfred , 52, 62
Eine kleine Nachtmusik, 53
Einstein, Albert , 67
electrical energy, 5
electrochemical processes, 6
electronic communications, 107

energy, 11, 16, 17, 29, 31, 49, 60, 61, 62, 110, 112, 114
England, 109
English language, 40, 41, 42, 43
entropy, 12, 17, 18, 23, 58, 59, 81, 82, 110, 111
enzymatic activity, 115
enzyme, 115
Eternal
 Dimensions, 80
 Mind, 55, 71
 Thought, 70, 71
Eternity, 71
European, 53
Evolution, 1, 2, 23, 24, 30, 52, 60, 61, 82, 85
 of life, 85
Evolutionary
 concept, 21, 107
 materialism, 22
 relationships, 100
 theory, 7, 30, 93, 99, 100, 107
evolutionists, 99
Extra Terrestrial Intelligence, 109
Faure, 95
fax machine, 71
Federal Institute for Informatics, 46
fire principle, 36
Flood, Raymond , 69
flood, 102, 103, 105, 106
forest, 102, 103, 105
fossil forest, 103, 105
fossil, 98, 101, 102, 103
Fox, 5, 6, 25, 54, 111
fundamentalist, 21
Gale, 95
Genesis report, 71
Genesis, 3, 91
genetic
 code, 40, 41, 51, 52, 54, 59, 65
 information, 50
 molecule, 65
Geological Survey of Canada,, 101
German, 42
Gillette, David D., 100
Gitt, Werner , 45
God and the New Physics, 71
God, 63, 70, 71, 75, 76
gold, 35
Gould, Stephan J., 1
graph, 68
greenhouse effect, 104
Hadrosaur bones, 97
half-life, 86, 87
half–toning, 39
hand–glove fit, 16
heart, 48, 56
heaven, 72, 75
Heiberg, Axel, 105
hell, 72
homo sapiens, 1, 22, 23, 107
human cerebrum, 65, 75, 82
human visage, 49
Hume's soul, 73
Hutchinson, 95
hypocycle, 62, 65
Iceland, 90, 96
information, 7, 8, 16, 17, 19, 23, 25, 26, 27, 28, 29, 36, 38, 39, 46, 47, 63, 82
 storage, 81
 theory, 36, 37, 38, 45, 48, 51
inherent properties, 7
inorganic, 54
 matter, 9, 22, 24, 28, 56, 60,

76, 110
universe, 56
instruments, 92
intellectual capacity, 66
International Symposium at Jülich, 26
Barrow, Isaac, 72
Newton, Isaac, 72
isotope, 89, 93
methods, 95
composition, 95
J. Monod, 52, 58, 62, 77
Jülich International Symposium, 30
keyboard, 52, 53
key–lock, 112
kidney, 39, 43, 48, 56
Kornberg, Arthur, 27, 110
language, 40, 42, 43
Laureate, Nobel 44, 49
lava, 91
Lavoisier, 35, 36
lead isotope, 95
lead, 35, 86, 87
Leibnitz, 73, 74
Leiden University, 98, 100
life's synthesis, 17
life, 6, 10, 16,19, 22, 25, 27, 28, 52, 81, 82, 112
Little Green Men, 109
Lockwood, Michael, 69
logos, 40, 63, 76
Lovell, 109
lung, 39, 56
machine, 6, 50, 55, 57, 60, 62, 65, 75, 76, 82
Marxism, 51
material universe, 56
Materialism, 77
materialist position, 71
Materialists, 72, 77, 80
matter, 8, 10, 57, 59, 62, 75, 78, 79, 82, 85, 108, 111
mechanism of thought processes, 48
metabolic machine, 28, 62
metabolism, 16, 17, 23, 113
metal calxes, 35
metallic oxides, 35
meteorites, 95
methane, 5, 111
Middle West State University, 88
Middle West, 86
Miller, 5, 6, 25, 54, 111
Mind of a Creator, 51, 70
mirror image, 15, 16, 23, 110, 111
modified by chance, 60
molecular fit, 11
Morse code, 109
mortal human, 65
Mozart, Wolfgang Amadeus, 53, 80
Mutation, 1
Muyzer, 100
Muyzer, Gerard , 98
mystic machines, 61
NASA, 109
natural law, 54, 56
natural selection, 1, 24, 26, 59, 60, 82, 83
Naturalistic
processes, 10, 50
chemistry, 19, 82
evolution, 22
Nature, 88
of Time, 69
of space, 78
negative age, 95

Neoplatonists, 75
Neo–Darwinism, 59
Netherlands, 98
New Mexico, 100
New York, 68
Noah's Flood, 103
Noah, 106
North American States, 53
North Pole, 102, 103
North Slope, 97
Northern Canada, 101
old earthers, 91, 92
optical activity, 23, 24, 30, 54, 59, 81, 110, 114
optical isomerism, 16
optical isomers, 110
optical purity, 10, 17, 23, 24, 25, 26, 27, 54, 58
ore, 87
origin of life, 8, 9, 23, 30, 32, 40, 51, 54
osteocalcin, 98, 100
outer space, 109
Oxford Union, 19, 22
  Debate, 19, 26
oxygen, 35, 36
Pacific, 104, 105
Pasteur, Louis , 27, 82, 109
Patterson, 95
Person, 57
personal Creator, 57, 59
Personality, 55
phlogiston, 36
physical shapes, 28
physical world, 75
planet earth, 105
plant world, 101
polar desert, 102
polarized light, 12, 16
Popper, Sir Karl, 63, 72, 99
potassium salt crystals, 90
Potassium/Argon method, 89, 90, 93, 96
potential information, 32, 52, 53
Priestley, 36
primate brain, 76
primate cerebrum, 62, 63
primate type, 61
primeval
  cell, 1
  chemical soup, 50
  forest, 105
  slime, 82, 85
printing processes, 39
protein, 97, 98, 99
pseudo scientific, 96
psychospace, 78, 79, 80
Radioactive
  elements, 88
  ores, 87
radiometric dating, 85, 86, 88, 91, 97, 98, 99
radiotelephony, 108
radium, 86, 87, 88
random processes, 23, 58, 59, 82
randomness, 7
Redwood trees, 106
Robbins, Lisa , 98
Rolls Royce, 71
Saint Augustine, 38
San Ysidro, 100
Sauropoda Diplodocidae, 100
scientific mythology, 92
Scripture, 103
Second Law of Thermodynamics, 57, 81
Seismosaurus halli, 100
self–winding watch, 81

SETI, 109
Shannon's bit of information, 43, 44, 45, 47, 50, 61, 77
Shannon's surprise effects, 53
Shannon, Claude , 17, 27, 31, 32, 35, 36, 37, 38, 42, 46, 110
Siemens Journal, 46
Sir Isaac Newton, 74
six days of creation, 71
Smith, Maynard , 21
Southern Australia, 104
space and time dimension, 49, 55, 72
space and time, 67, 70, 71, 74, 76, 79, 80, 108
space, 73, 74, 78
Space–Time Continuum, 67, 68, 70, 71, 72, 78
spatial arrangements, 45
Spiegelman, Sol , 28, 110
spontaneous generation, 8, 22, 24, 82, 109, 110
spontaneous naturalistic stochastic chemistry, 44
spontaneous processes, 19
spontaneous stochastic natural chemical processes, 44
spontaneous synthesis, 6
steam engine., 37
steam, 5
stochastically produced surprise effects, 44
stochastically synthesized, 41
surprise effect, 42, 44, 45, 48, 51, 52, 58, 61, 62, 67, 76
Surtsey, 90, 94, 96
Tampa, 98
tectonic activity, 106
teleonomic concepts,, 51
teleonomic information, 10, 32
teleonomic life, 51
teleonomic organs, 56
teleonomy, 9, 10, 47, 49
television, 78, 80
theistic evolution, 60
Thermodynamics, 9, 80
thinking machine, 63, 65
thorium, 88
thought material, 79
thought processing machine, 65
thought, 78, 79
time dimension, 67, 70, 71, 108
time spans, 107
time, 10, 29, 31, 33, 38, 53, 62, 67, 72, 73, 74, 76, 80, 85, 88
timeless eternity, 70
tree rings, 94
Trinitarian Person, 55
Tudge, Paul , 101
units of information, 39
universal think–tank, 64
universe, 96
University of Göttingen, 62
University of Oxford, 20, 22
University of South Florida, 98
uranium, 89
Uranium/Lead, 93
von Neumann machine, 7, 9 31
Washington, 68
Western Academe, 51
Wilder–Smith, A. E., 20
Williams, Alexander R., 95, 96
word processing, 40
Word, 63
world–line, 67, 68

Books by
Dr. A.E. Wilder–Smith

TWFT Publishers
P.O. Box 8000
Costa Mesa, Ca. 92628
1 (800) 272–WORD

AIDS: Fact without Fiction________________ Item# AW001

A comprehensive, up-to-date presentation of the facts about this new plague. Discusses its discovery, its spreading about the world, its medical effects and the efforts to overcome the disease as well as the politics involved.

Why Does God Allow It? _________________ Item# AW002

If there is a God — why does He permit all the violence and suffering in the world? Sensible answers to questions that have plagued people since the beginning of time.

He Who Thinks Has To Believe ____________ Item# AW003

A hypothetical story about a society of "Primitive" Neanderthal people discovered by a modern rescue team. The rescuers are amazed to find that the natives have quite reasonably come to the conclusion that there is a Creator God.

The Natural Sciences Know Nothing of Evolution _____________________ Item# AW004

The Natural Sciences themselves are contrary to the theory of Evolution as is demonstrated in detail in this challenging thesis.

The Scientific Alternative to Neo–Darwinian Evolutionary Theory _____________________ Item# AW005

Modern Darwinism is in a constant state of flux. New theories are developed almost daily to try to explain away the obvious difficulties presented by advancing science, often casting off reason and experimental verification in preference to postulating the existence of a Creator. Here we present the scientific alternative to Neo–Darwinism.

The Creation of Life______________________ Item# AW006

A classic, exposing scientific materialism and concluding that to have an efficient design you must have an efficient designer.

Is This a God of Love? _________________ Item# AW007

Many today stumble at the seeming conflict between the truth that God is Love, and the obvious suffering we see in the world around us. This book is primarily written to the unbeliever who wants to believe, but lacks a good perspective of the world in which we live.

Man's Origin, Man's Destiny _____________ Item# AW008

In recent history mankind has undergone a complete reversal of perspective on the origins of the human race from one of belief in the supernatural to another which proposes a purely naturalistic explanation. The inevitable result is that men no longer have a clear sense of purpose; of the meaning and sanctity of life. Only by understanding the origin of all things can we hope to regain our focus on who we are now and what is our eternal destiny.

Baptism and its Influence on
Christian Devotion ____________________ Item# AW009

Opinions on the importance of Baptism today range from indifference to vital for salvation. This balanced approach dispels some of the misconceptions and discusses the importance of obedience to the command of the Lord Jesus to, "Go ye therefore, and teach all nations, baptizing them in the name of the Father, and of the Son, and of the Holy Ghost…" Matt 28:19.

The Time Dimension: Its ________________ Item# AW010
Relationship to the Origin of Life

Today's science is noted for quoting exceedingly long time spans for the history of the proposed evolution of life on earth. In this book, Dr. Wilder–Smith challenges such thinking on the basis of some of the latest developments in information theory, quantum physics and radiometric dating.

The Day Nazi Germany Died ______________Item# BW001

An autobiography by Beate Wilder-Smith, Dr. Wilder-Smith's wife. An eyewitness account of life in Nazi Germany and the Russian and Allied invasion.

Dr. A.E. Wilder–Smith